Car Stereo Speaker Projects Illustrated

Car Stereo Speaker Projects Illustrated

Daniel L. Ferguson

McGraw-Hill

New York   San Francisco   Washington, D.C.   Auckland   Bogotá
Caracas   Lisbon   London   Madrid   Mexico City   Milan
Montreal   New Delhi   San Juan   Singapore
Sydney   Tokyo   Toronto

Library of Congress Cataloging-in-Publication Data
Ferguson, Daniel L.
    Car stereo speaker projects illustrated / Daniel L. Ferguson.
        p. cm.
    ISBN 0-07-135968-0
    1. Automobiles—Audio equipment.   2. Loudspeakers.   I. Title.
TK7881.85.F47 2000
629.2'77—dc21   00-038009

McGraw-Hill

A Division of The McGraw·Hill Companies

456789 QWC 876

ISBN 0-07-135968-0
ISBN 13 978-0-07-135968-9

The sponsoring editor for this book was Scott Grillo, the editing supervisor was Frank Kotowski, Jr., and the production supervisor was Sherri Souffrance.

It was set in Palatino by North Market Street Graphics.

Printed and bound by Quebecor World Bogotá

McGraw-Hill books are available at special quantity discounts to use as premiums and sales promotions, or for use in corporate training programs. For more information, please write to the Director of Special Sales, Professional Publishing, McGraw-Hill, Two Penn Plaza, New York, NY 10121-2298. Or contact your local bookstore.

This book is dedicated to Pat,
my beloved wife of 33 years,
for all of her assistance and encouragement
throughout this project.

Contents

Introduction

In general, loudspeakers are the weak link in the sound reproduction equipment chain. And while today's commercially produced speaker systems have steadily improved in price and value, consumer electronics continue to advance at a faster pace than loudspeakers. Designing and building them continues to be a combination of art and science. So for the hobbyist, this means that there is still room to experiment with building your own speakers—and maybe even save some money while you're having fun creating your own designs.

A real opportunity area for amateur speakerbuilding continues to be car stereo. I'm sure most of you reading this could agree that car manufacturers are still not installing good quality speaker components as original equipment, and that speaker locations in many cars appear to be questionable. The best proof of all of this is that the after-market car stereo industry continues to thrive year after year. Somebody out there must be looking for something better.

Over the past 15 years, I have had the opportunity to build and experiment with a variety of car stereo speaker systems. Prior to building that first one, I read many books and articles on speaker-building to try to understand the principles. At that time, there were no books specifically written for car audio; they were all geared towards home speaker construction, so application to car audio was by inference. When I ventured out into the auto sound world of 1985, I found few reasonably priced, commercial speaker systems that I really liked. If I liked the way a system sounded, it was invariably too expensive. A simple two-way, box system with 8-in woofers from some of the major manufacturers cost the equivalent of $1,000 or more in today's dollars. Also, I was more than dismayed by some of the concepts being used at that time—many of which have been mercifully abandoned.

So out of necessity, I decided to try my hand at building my own car stereo speaker system—for financial reasons and out of general dissatisfaction. Over time, as my personal vehicle changed, it afforded me the opportunity to experiment with different concepts. As others saw and heard the results, they were moved to build systems similar to mine or contract me to build one for them.

Eventually, I reached the point where I felt I had enough unique material to write a book on the subject. The result was *Killer Car Stereo on a Budget*, which outlined all of the concepts I believed to be improvements over the commercial systems of that period. I was fortunate enough to get the attention of Audio Amateur Corporation, which published it. Seven years later I wrote a much improved sequel, *Ultimate Auto Sound*, after more experimentation and the benefit of a personal computer.

That brings us to today and the purpose of this book: To show you in step-by-step photos and illustrations how to build a variety of car stereo speaker systems using designs and concepts which have been proven to be extremely reliable. You should be able to do the work and reproduce the results even if you've never built anything like this before. Since much of this involves semiskilled woodworking, I will also show you how to accomplish this with minimal investment in tools and equipment. By following simple directions, you should be able to construct your own car stereo speaker systems that have both superior performance and reasonable cost. As an added benefit, you will probably learn enough to apply these principles to other applications, like home audio, and have even more fun in the process.

Concepts

All of the systems I advocate for use in cars employ a combination of high-quality satellite speakers and one or more powered sub-woofers. This in itself is certainly not news in today's world of audio. If anything, it is the norm. What you will see here, however, are unique ways to apply these concepts in order to create systems that fit in limited spaces and generally perform better than commercial systems. Even if you wanted to purchase something similar, many of these designs don't exist commercially. And if they did, it's doubtful they would custom-fit your particular application. For the most part, you should be able to do what professional installers do and get better results since you will be applying different principles and can take as much time as you want.

The speaker projects contained in this book will greatly enhance the sound of almost any head unit (radio). While not directly related to speakerbuilding, another somewhat unique concept I advocate is to retain the use of the factory head unit whenever possible. In general, I have found them to be attractive, rugged, and reliable. Many of today's units sound great and look like they belong in your particular vehicle. Best of all, you already own it. Unless there's a known deficiency, like no CD options, plan to at least try to retain the factory head unit.

The subwoofer designs I advocate employ a somewhat unique subwoofer filter, which I developed back in the 1980s and have improved over the years. To my ear and many others, it just plain sounds better. I will show you in great detail how to construct it yourself for a modest cost. Alternatively, if you'd rather not take this on, you can elect to purchase a commercial version from me direct.

In general, this book is not about high sound pressure levels, although virtually every one of the systems we'll build is capable of inflicting hearing damage if used improperly. While we won't be building very large systems as examples, there will be plenty of information provided to enable you to do so.

Woodworking can be fun or maddening, depending on whether or not you have the right tools to do the job. To minimize the frustration, all of the projects we will build will require the minimum in the way of tools in order to make this process both fun and affordable. To make it fun, we'll employ some tips and tricks to get around the limitations of the simple tools we'll be using.

Component selection is always a major factor in the success of a speakerbuilding project. Some of the bass drivers (woofers) we will use are unique in their ability to provide low-frequency extension in small enclosures. While it's fairly easy to get big sound from big enclosures, it took quite a bit of experimentation to get similar results from small ones. You will get the benefits of all the work done in the past.

Cost is a major design consideration for every project in this book. The particular components and electronics we advocate using are selected to provide good performance and reasonable cost.

This book is divided into sections that show the application of my designs in various positions and vehicle types. For many of the examples constructed, there will be tables that give all the parameters to permit you to build larger or smaller versions so that you can custom build something that fits your particular application.

For the skeptics out there (I'm one of them), the performance of every project we build will be authenticated using an Audio Control SA-3055 real-time analyzer (RTA), graciously loaned to us by the manufacturer for the duration of this project. A picture of this terrific analytical tool is shown as Photo I-1. In addition, we will use oscilloscope displays and other instruments to show you actual data wherever possible.

Finally, lets be clear about what this book is not. It's *not* a car stereo installation manual. It is a construction project book for car stereo

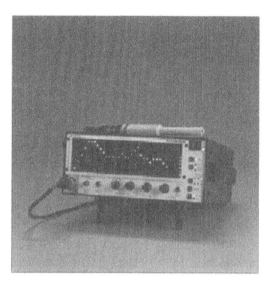

Photo I-1 Audio
Control SA-3055
Real-Time Analyzer.

speakers. The installation part of the job has been covered by others
(including myself in past work). This book is intended to let you see
first-hand how to go about the actual construction of car speaker systems.

Having laid out the ground rules, let's go to work!

Daniel L. Ferguson

Acknowledgments

I would like to express my sincere thanks to the Audio Control Corporation for the loan of their outstanding SA-3055 Real Time Spectrum Analyzer, which made it possible to authenticate each of the projects we constructed.

Many thanks also to Larry Hitch and Madisound for providing many of the speakers and components used in the construction projects.

Chapter 1

Front Speaker Projects

In this first section, we'll tackle some of the problems you might encounter with installation of the all-important front speakers. These provide over 90 percent of the musical information and determine to a large extent the overall quality of the system. While they are usually not the most expensive component in the system, they are the most difficult to get right. This is primarily due to the limitations of the speaker placements available.

For instance, I've listened to some very popular foreign and domestic vehicles that could not produce any semblance of a front sound stage unless I drastically altered the right-to-left balance. In these instances, leaving the balance control in the centered position caused either the driver's or passenger's speaker to be very prominent—depending on the particular vehicle. In some of these cases, the acoustic balance could only be corrected by moving the balance control 30 or 40 percent off center. While this may be tolerable for a single occupant, it would be unnerving for anyone else in the vehicle.

If audio is one of your priorities, the time to check for a front sound stage is before you buy the vehicle. But let's face it, many circumstances cause this to be lower on the priority list when we're shopping for that new car or truck. So now, for whatever reason, you are dealing with a front sound stage (or "imaging") problem. One cause of poor imaging is poor off-axis response from the front speakers. To correct this, I always start by replacing the factory speakers with the best-sounding aftermarket coaxes I can find to determine if the stock locations have the potential to produce a reasonable sound stage. The hope is that if this is successful, no alteration to the vehicle's interior will be required.

If you're wondering how replacing the stock speakers could help with imaging, it has to do with the fact that car makers don't usually install coaxes with separate tweeters as original equipment. Since a 5- or 6-in cone speaker has very limited dispersion (off-axis response) above 2,000 or 3,000 Hz, a significant portion of the spectrum is probably being directed in a concentrated beam to somewhere other than your ears. This is compounded by the fact that higher frequencies are very directional. So installing a high-quality coax usually means we are adding a tweeter where one did not exist before, which has the potential to fill in some gaps in the audible spectrum.

Another possible fix for an imaging problem is to install additional speakers in the front doors—a major decision. Any alteration of your car's interior should be approached with quite a bit of caution. If you make a mistake, the consequences could get expensive. But a case can be made for improving the on-axis and off-axis response versus distance combination—especially if the car manufacturer left you with no other viable alternatives.

Maybe you have a simpler need—you just want more midbass up front and the puny 3½-in dash speakers just won't do it by themselves. For whatever reason, you've arrived at the point that you are ready to install additional speakers in your door panels. In this section, we're going to look at some things we can build to facilitate or even improve this installation. Let's start with something simple.

Spacer Rings

Let's assume that you have decided to attempt to improve your front sound stage by installing some aftermarket coaxes in the door panels. Photo 1-1 shows a typical high-powered 5-in coax. This one

Photo 1-1

happens to be a Madisound 5402 with a measured frequency response shown as Figure 1-1. It is equipped with an Audax ½-in dome tweeter for extended highs. The woofer has a large magnet for good power handling and a polypropylene cone for maximum moisture resistance. The large magnet structure can be a problem if it extends too far into the door cavity. The fix for this is to install a spacer ring between it and the door panel.

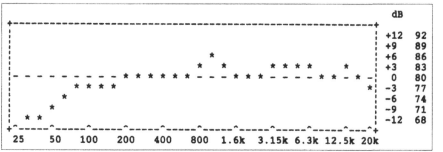

Figure 1-1
Frequency response of Madisound 5402 coax measured in 4- by 8-ft test wall.

In order to prepare for this installation, you will have to study the problem carefully and decide on exactly the position where you want to install the new door speakers. Then, you must remove the door panels to see what is behind them.

There are two aspects of the door mounting problem. One is the placement of the magnet, and the other is locating enough structural steel to get a minimum of two of the four speaker mounting screws into. Let's assume that, after careful examination, you have determined that the new speaker will fit perfectly except the magnet sticks out too far and will either hit a structural member or interfere with the window mechanism. So now you need some type of spacer to fit between the speaker and the door panel.

In this section we'll show you how to build two different types of spacer rings. We'll start out with the simplest tools first and move up from there. The second type, using more power tools, should yield a more attractive installation.

Simple Spacer Ring

The tools required are the following:

- Electric drill
- Saber saw
- Wood rasp
- Compass
- Ruler
- 100-grit sandpaper
- Work stand (optional)
- C-clamp (optional)

For materials we'll need the following:

- High-quality particle board (best source for this for small projects is precut shelving available at home improvement stores like Lowe's)
- Water-based wood putty (e.g., Elmer's, Durham's)
- Suitable spray paint (I prefer the black satin spray enamel available at Wal-Mart for about $1.50 a can)

Procedure

1. Lay out the spacer by drawing two concentric circles. The outer circle should be slightly larger than the grille by $\frac{1}{16}$ to $\frac{1}{8}$ in. (It looks fairly ugly if the spacer is undersized.) The inner circle is the diameter of the woofer mounting hole. Measure the outer diameter of the grille as shown in Photo 1-2, add $\frac{1}{16}$ in or so to the radius, and draw the outer circle on the particle board as shown in Photo 1-3.

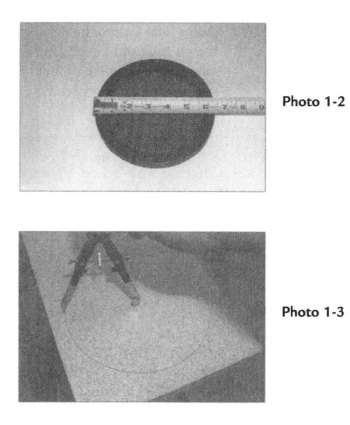

Photo 1-2

Photo 1-3

2. Carefully measure the inner diameter of the speaker to determine the size of the mounting hole and draw the inner circle. Photo 1-4 shows one technique for doing this accurately, which involves using a framing square and speed square to form a caliper. Use this dimension to draw the inner circle.

Photo 1-4

3. Drill a hole (approximately ⅜ in) inside the inner diameter as shown in Photo 1-5 and cut out the inner diameter with the saber saw as shown in Photo 1-6. Test-fit the coax and correct any problems before proceeding.

Photo 1-5

Photo 1-6

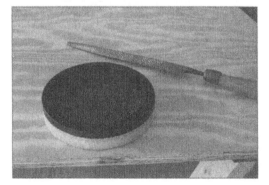

4. Cut out the outer diameter (or traced outline if the grille is noncircular) with the saber saw.
5. Test-fit the grille to the spacer and correct any irregularities with the rasp as shown in Photo 1-7.

Photo 1-7

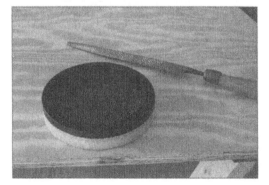

6. When everything fits properly, fill the particle board edges with water-based putty. Add a little water to thin the consistency, and rub the mixture into the spacer edges with your finger or a damp rag. Let dry overnight and sand smooth. The results are shown in Photo 1-8.

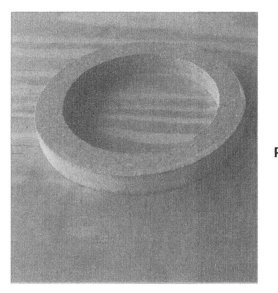

Photo 1-8

7. Paint the spacer as appropriate, and it is ready to install as shown in Photo 1-9.
8. Installation is accomplished by running screws through the woofer, grille ring, and spacer into the door frame. As stated earlier, all of this must be laid out carefully in advance.

Photo 1-9

Contoured Spacer Ring

While the appearance of the simple spacer is OK, if we're objective about it, we might conclude that it could look better. For one thing, it looks exactly like what it is—a spacer. It could look a lot more like a finished part with a few improvements. For example, if space permitted, it would look considerably better if it were a little larger than the grille so that the hole could be recessed slightly. Also a rounded-over edge should soften the look and make it more appealing. Photo 1-10 shows how this could look. However, to make this spacer, we'll need a plunge router. For this job, we'll be using a straight ⅜-in-diameter carbide-tipped bit and ½-in-radius roundover bit as shown in Photo 1-11.

Photo 1-10

In order to cut perfect circles with the router, we'll need to make a simple fixture from a piece of scrap particle board as shown in Photo 1-12. The actual dimensions are noncritical, but material thickness should be ½ to ⅝ in. For this application, I chose a convenient piece of scrap that happened to be 8 by 12 by ⅝ in. To make the fixture, trace the router baseplate mounting holes (they all have them) onto the scrap at

Photo 1-11

a point that positions the router an inch or two from one end. Drill and countersink the holes and secure the fixture to the router with some suitable flathead bolts and nuts. Be sure the bolt heads do not protrude! If you haven't already done so, install the straight bit in the router and plunge the bit through the fixture. Retract the bit. The edge of the hole now defines the edge of all future circles cut with the fixture.

Photo 1-12

To set the radius of the circle to be cut, mark the center on the bottom of the fixture. Measure from the outer edge of the hole back toward the long dimension of the fixture as shown in Photo 1-13. Drill a small hole and drive in a snug-fitting nail with the head removed as shown in Photo 1-14. For this spacer there will be a total of four cuts; therefore, you may want to measure and predrill them all at this time. Figure 1-2 shows the dimensions required for the nominal 5-in stamped steel grille supplied by Madisound. It is typical of many others. However, the grille you are using may differ significantly and will require close measurement to make the spacer fit for the best appearance.

Photo 1-13

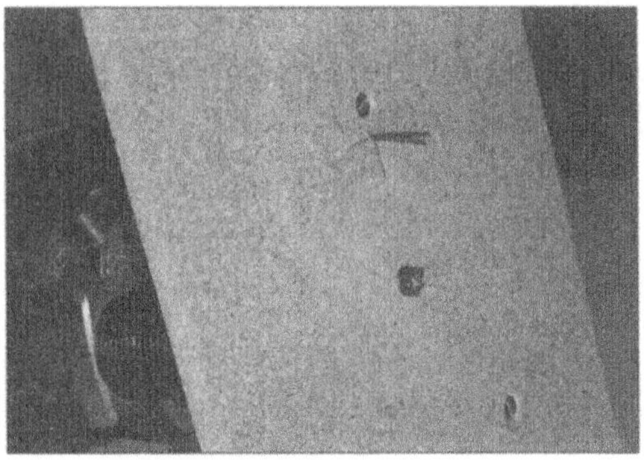

Photo 1-14

Figure 1-2 Routing diagram for contoured spacer ring.

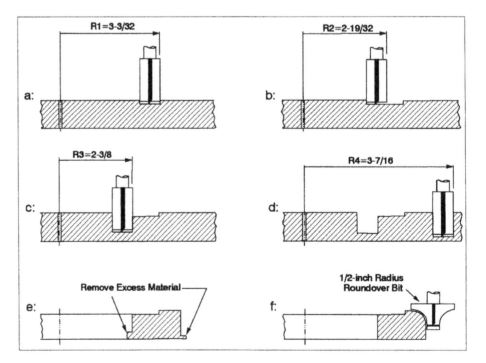

When you're satisfied that the center-pin holes are correctly located, test cut all four radii in a piece of scrap. Adjust center-pin positions as necessary until you get the correct results.

Procedure

1. When all test cuts are at the right diameters, set the fixture up on a fresh piece of material and begin by cutting the recess for the grille. This could also be referred to as a counterbore or rabbet and will require two passes to get a wide enough land area to fully support the grille. This is a very shallow cut—$\frac{1}{32}$ to $\frac{1}{16}$ in deep. The first pass is shown as Figure 1-2a. Widen the rabbeted area by decreasing the radius and making the second cut shown as Figure 1-2b.

2. Remove the router fixture, clean out all the sawdust, and test-fit the grille in the recess. If it looks OK, reposition the nail to the inside radius and make a cut to a depth of about one-third of the way through. Remove the fixture, clean out the sawdust, and reposition the fixture. Repeat this one more time until the depth of the cut is approximately two-thirds of the way through as shown in Figure 1-2c.

3. Remove the fixture and reposition the nail for the outside diameter. Make this cut in three or four passes as shown in Figure 1-2d. On the final pass, take a position that allows seeing the back side of the workpiece (or use a mirror). Plunge the router all the way through and proceed with the final cut. Stop the router when approximately 15 percent of the material remains. Remove the router and cut through the remaining material with a saber saw.

4. Cut the center plug free with a saber saw after drilling a hole to get the blade started. At this point you will have the partially finished ring shown in Figure 1-2e. Carefully remove the remaining tab from the outside surface and the ridge from the inside bore by filing and sanding smooth.

5. The final step is to round over the outside edge as shown in Figure 1-2f. Remove the fixture from the router and replace the straight bit with the roundover bit. This should also be carbide-tipped and have a ball bearing guide. If you have a router table, this is a good application for it as it makes this operation a breeze. If you don't, you have two options. One is

to attach the spacer ring to a secure, flat surface by installing screws in the back side and making the roundover cut from the top. Alternately, you can improvise a router table by enlarging the hole in the circle cutting fixture to accommodate the roundover bit and then clamping it securely in an inverted position to a work stand as shown in Photo 1-15. Make the roundover cut in two passes to minimize the force on the ring. Again, the result is shown in Photo 1-10.

Photo 1-15

6. Rub wood putty into the outer edges and let dry overnight. Then, sand smooth and paint with several coats of semigloss black paint to match the grille. Alternatively, you could also paint this to match the color of the trim in your vehicle. The finished part is shown as Photo 1-16.

Photo 1-16

7. Mounting of this spacer ring is identical to that of the simple spacer ring, above.

I'm sure that by now you've figured out that the plunge router and jig we made are a great way to cut all kinds of holes and circles. For instance, we could have easily used it to cut out the simple spacer ring. It would have been much more accurate and looked a lot better.

Two-Way Component System

When it comes to door speaker systems, individual components are considered to be the ultimate. There are several reasons for this, but the biggest and most obvious one is that each of the components is usually of much higher quality than is found in most coaxes. In fact, the quality of this type of speaker system is really limited only by your budget. However, even if we use the highest-quality components, performance will still be greatly dependent on placement.

We can improve on the placement problem by designing in the widest off-axis response possible. Key parameters for this are: one—choosing a tweeter with good dispersion and two—setting the crossover frequency as low as possible to minimize "beaming" from the midbass driver (woofer). If there is sufficient space available, we can install the components on some type of rigid backing to reduce movement of the individual drivers, which causes a type of distortion referred to as "smearing." And while we're at it, we'll make all the corners and edges as smooth as possible to minimize diffraction.

So for the two-way component system we'll be constructing, we choose the best components we could find that met our design needs and are yet still cost-effective. With the help of the designers at Madisound, we selected the following:

Tweeters: LPG 26NA, 1-in aluminum dome.
Woofers: Vifa P13WH-104, cast aluminum frame, 5-in polypropylene cone with rubber surround, 4-ohm impedance.
Crossovers: Madisound two-way, third-order network with premium, polypropylene capacitors and air core inductors, designed using LEAP loudspeaker modeling program. The crossover schematic and wiring diagram are shown in Figures 1-3 and 1-4.

The components for one side are shown in Photo 1-17.

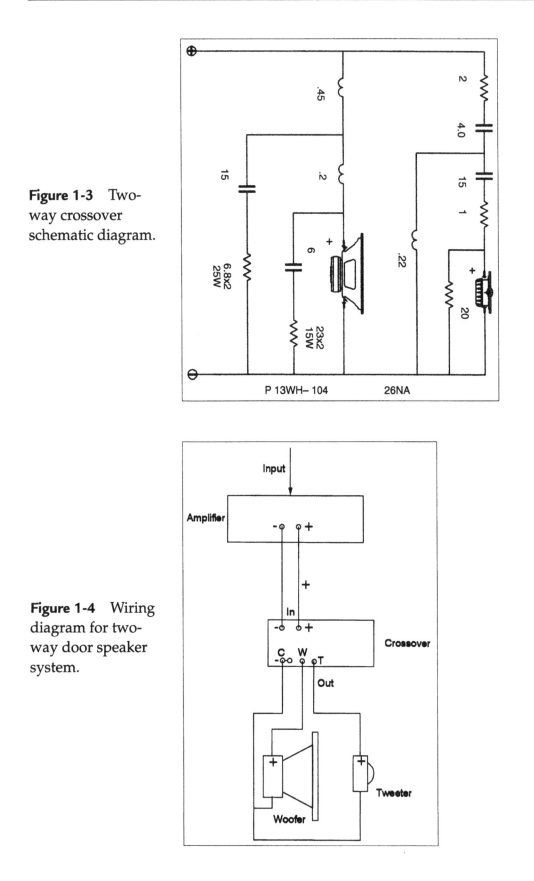

Figure 1-3 Two-way crossover schematic diagram.

Figure 1-4 Wiring diagram for two-way door speaker system.

Photo 1-17

We will start by designing a suitable mounting board to fit a generic door panel. We'll dress it up by recessing the components and rounding over the edges as was done for the contoured spacer. Lastly, we'll measure on- and off-axis system response on a flat panel to demonstrate its accuracy.

Notes:
1. If you prefer to not use a mounting board, you can mount the woofer and tweeter separately. The tweeter comes in several different configurations and can be flush- or surface-mounted. If you plan to install the tweeter flush-mounted, choose the flanged version.
2. To get predictable frequency response results, maintain spacing between the woofer and tweeter as close as possible to Figure 1-5.

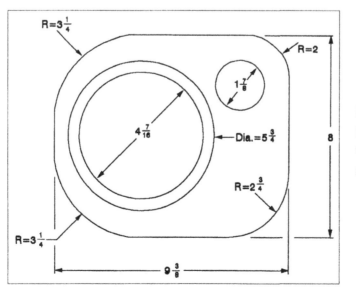

Figure 1-5 Mounting plate for two-way component system.

Procedure

1. Installation of this system will require you to cut a 4¾-in hole in your front door panels. And, as stated before, this is a major decision, which requires a great deal of study. You may also want to enlist the help of a professional installer before proceeding.

2. Assuming we've done all the prework, we begin by tracing a paper pattern of the door contour which will be transferred to a full-sized drawing of the driver's side as shown in Figure 1-5.

3. After tracing the pattern onto the particle board, we lay out the hole pattern for the woofer and tweeter with a compass.

4. Using the router circle cutting jig, we route the two mounting holes. Photo 1-18 shows the tweeter hole after the final router pass. The remaining material holding the plug in is cut with a saber saw and the hole is smoothed with the wood rasp as shown in Photo 1-19. The test fit of the tweeter is shown in Photo 1-20.

Photo 1-18

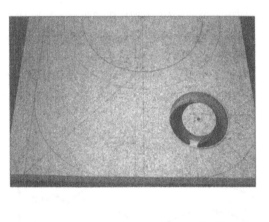

Photo 1-19

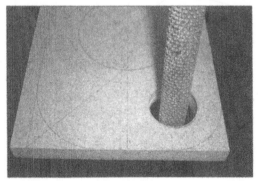

Photo 1-20

> *Note:* The tweeter mounting configuration we chose was strictly for appearance purposes. Boring the proper mounting hole for it requires much more precision than would be normally expected, and even then the tweeter must be held in place with silicone caulk. If you opt for the flanged version, mounting the tweeter is much less problematic.

5. Next, we cut the optional recess for the woofer and grille and test-fit the grille as shown in Photo 1-21.

Photo 1-21

6. The bore for the woofer is shown in Photo 1-22.

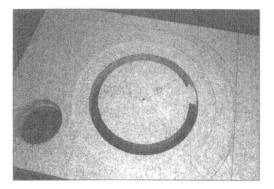

Photo 1-22

7. After confirming that the woofer and tweeter fit properly (Photo 1-23), we cut out the driver's side mounting board with the saber saw.

Photo 1-23

8. The passenger side mounting board will, of course, be the mirror image of the driver's side. After it is cut to size, place the two pieces together and correct any obvious differences by filing and sanding.

9. Using the router and the ½-in-radius roundover bit, round over the outside edges. The finished driver's side is shown in Photo 1-24. Fill the edges with thinned wood putty, and let dry overnight.

Photo 1-24

10. Sand the boards smooth, wipe with a tack rag, and apply several coats of paint. The finished driver's side is shown in Photo 1-25.

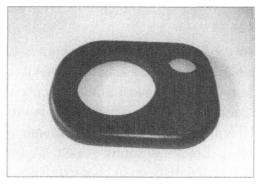

Photo 1-25

11. The mounting board can be best secured to the door panel with screws installed from the back side combined with running sheet-metal screws through the woofer and grille ring and into the metal door frame. All of this must of course be carefully taken into account during the initial planning phase. Failure to anchor the board to the door frame properly (at two or three strategic places) will probably result in an unsatisfactory installation. Photos 1-26 and 1-27 give a feel for what the installed system would look like.

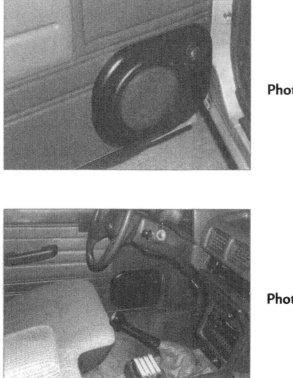

Photo 1-26

Photo 1-27

Testing

To confirm that we have a high-accuracy design we mounted the completed plate system on a heavily reinforced, ¾-in-thick plywood test wall and measured its response with the Audio Control SA-355 real-time analyzer (RTA) as shown in Photo 1-28. We compared the RTA plots to the predicted response curves generated at Madisound using LEAP—a sophisticated loudspeaker modeling program. To determine dispersion characteristics, we measured the response on-axis, at –30 degrees horizontal (left of center) and at –30 degrees horizontal combined with +30 degrees vertical to approximate the expected listening position in a typical installation.

Photo 1-28

Figure 1-6*a* is the response predicted by LEAP 30 degrees to the left of center. Directly below it is Figure 1-6*b*, which shows the response measured with the RTA at a distance of 18 in also 30 degrees left of center. The dip between 5 and 8 kHz was the biggest discrepancy between the two plots. However, when the mike was repositioned to 30 degrees above center, combined with 30 degrees to the left, the response shape became more similar to the model. One problem encountered was that the tweeter level (volume) was lower than it should have been, so I placed a jumper across the 1-ohm series resistor and achieved the response shown as Figure 1-6*c*. The tweeter level is still a little lower than ideal, but it looks fantastic considering how far off-axis the measurement is being taken and is actually quite uniform—a credit to LEAP and the designers at Madisound.

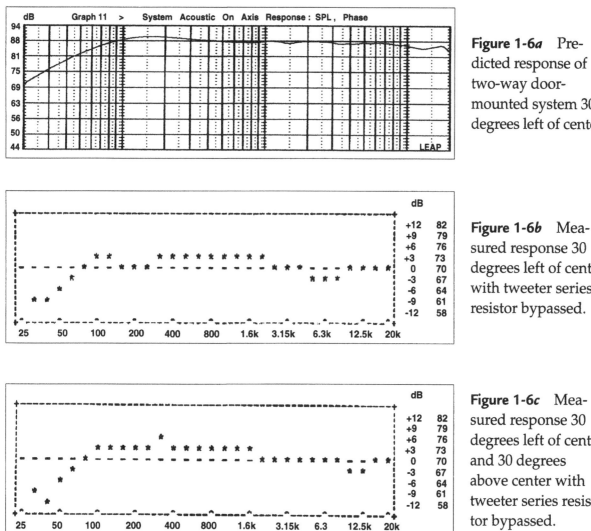

Figure 1-6a Predicted response of two-way door-mounted system 30 degrees left of center.

Figure 1-6b Measured response 30 degrees left of center with tweeter series resistor bypassed.

Figure 1-6c Measured response 30 degrees left of center and 30 degrees above center with tweeter series resistor bypassed.

So, how do they sound? In a word quite good, as I would expect from such a uniform frequency response curve. The word I would use to describe them is "smooth." They have a very wide listening window which, of course, was one of our main design objectives. Also, there was no trace of harshness in the highs, which is characteristic of some aluminum dome tweeters.

To get the full effect, I teamed them up with a good-quality subwoofer and the results were totally seamless. The illusion was that all of the bass was being generated by the Vifas due to their solid midbass capability. All in all, I am satisfied with the result.

While our system was designed by professionals using computer-aided design tools, we amateurs can do a pretty fair job of designing our own. *Speakerbuilder* magazine has been in publication since 1980 and is devoted solely to this hobby, so there is still considerable room for the hobbyist to experiment—even if you only have minimal test equipment. In Chapter 8, we'll demonstrate some of the basics of speakerbuilding.

That about wraps up this chapter on front speaker projects, although volumes could literally be written on dealing with this extremely difficult problem. I hope in this somewhat brief attempt that I have given you some ideas that you will find useful for tuning up your front sound stage.

Chapter 2

Rear Speaker Systems for Sedans

After the front speakers, the next most difficult speaker system to implement properly is the rear speakers in a sedan. We have not one but two systems to install in this location—the rear mains and the subwoofers. We will make the assumption that available space is limited to the package shelf under the rear window. The challenge—and it's a big one—is to do all this while utilizing the existing factory cutouts.

Now there are many car stereo enthusiasts out there who think that an acceptable solution to the sedan subwoofer dilemma is to simply place a large box system in the trunk with no direct coupling to the passenger compartment. There are others who have cut a hole in the trunk wall and installed a large-diameter subwoofer which fires through the rear seat structure and cushion. I have a question for those of you who subscribe to these types of installations. In a home system, what audiophile would place a subwoofer in the next room? Or, what music lover would bolt a subwoofer to the back of a sofa? The answer is, no one would—for obvious reasons. These practices have become accepted over time into car audio because manufacturers have offered few viable solutions to the sedan problem. In all likelihood, most of this evolved from just plain expediency.

What's wrong with all the above is that the whole principle behind a loudspeaker is to couple the motion of the moving diaphragm to our ears through the medium of air. When anything else gets in the way, like steel, fiberboard, insulation, thick foam rubber, etc., the sound is compromised, to say the least. Distortion is introduced in a big way and frequency response is altered.

Another serious side effect is the damage a large, uncoupled subwoofer box system can do to a car's structure from inside the trunk. I've seen interior trim screws back out, and I've personally observed trunk lids flexing with every beat of the kick drum. I also invariably hear loud rattles and buzzes intruding into the music(?).

For all of the above reasons, we will confine ourselves to building rear speaker systems that are directly coupled to the passenger compartment air space. They should have lower distortion, be more efficient, and be less destructive. Maybe best of all, you will still have a trunk when we're finished with each of the three designs explored in this chapter.

The Triax

Sedan System No. 1 is based on the assumption that your vehicle is equipped with 6- by 9-in oval rear speakers as original factory equipment, which is very common for many domestic cars. For this situation, we will fabricate our own version of a triaxial speaker system. The general arrangement is shown in Figure 2-1. It will employ a JVC CS-HX420 4-in coax to cover everything but the bottom two octaves. These particular coaxes were selected because of their smooth sound and compact size. In this application, power demands on the coaxes will be minimal since they only have to provide rear fill. Bass frequencies will be handled by an 8-in Madisound (Resource 1) 81524-DVC woofer, which has dual 4-ohm voice coils. To ensure plenty of headroom, these woofers have a relatively high 5-mm peak-to-peak excursion limit (referred to as X_{max}). In this case, we will be using only one of the voice coils (a somewhat unusual configuration), which will yield a 4-ohm impedance and the correct Q (related to damping) for this "free-air" trunk-mounted setup. The two drivers used in this project are shown in Photo 2-1.

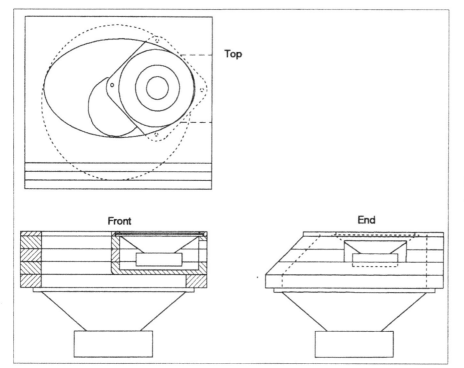

Figure 2-1 Sedan triax arrangement.

Photo 2-1

Looking ahead, the predicted subwoofer frequency response, when using this driver, is shown in Figure 2-6a (page 39). Other high-quality woofers may certainly be used in this application provided that their parameters are suited to free-air mounting and they have sufficient power handling and cone travel. As a point of interest, the free-air resonance frequency f_s for this particular sample measured 31.8 Hz, which is within 0.1 Hz of the published specifications. Driving only one of its two voice coils increased its total Q (Q_{TS}) from the specified value of 0.31 to 0.54, making it nearly ideal for this free-air application.

Getting Started

The triaxes are built on a series of adapter boards which couple the various drivers to the passenger compartment through the single opening. Figure 2-2 shows the details. While it appears to be complex, each piece taken separately is not.

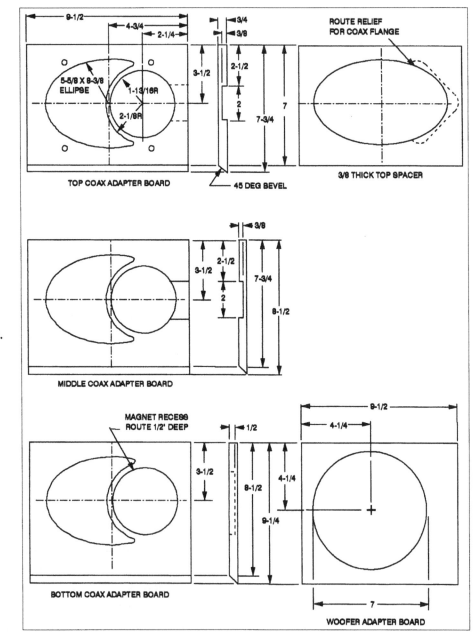

Figure 2-2 Sedan triax adapter details.

Tools required for this project are:

- Circular saw
- Saber saw
- Electric drill
- Router (optional)
- Table saw (optional)
- Wood rasp
- Square
- Clamp-on straightedge (recommended)
- Ruler, tape measure

Materials required for this project are:

- High-density particle board, ¾ in thick
  Note: A convenient source for this is 11½-in-wide precut shelving, available at building supply house like Lowes in 8-ft lengths. Another possibility is 24-in-wide countertop material available at cabinetmaker supply houses.
- Yellow carpenter's glue
- Water-based wood putty
- Sandpaper
- Screws:
     24—No. 6 flathead sheet-metal, ¾ in long
     6—No. 6 panhead sheet-metal, ¾ in long
     36—No. 8 panhead sheet-metal, 1¼ in long
     16—No. 8 panhead sheet-metal, 1 in long

Before proceeding at full speed with this project, test to see if the units will fit your car. Start by cutting out the top coax adapter board to its overall dimensions and attempt to fit this under the package shelf inside the trunk. If it fits and you are sure you will be able to attain an airtight seal with the steel package shelf, proceed with the rest of the fabrication.

Procedure

1. Begin by ripping a suitable length of ¾-in-thick, high-density particle board to a width of 9½ in. This is best done on a table saw but can also be done using a circular saw and straight-

edge (Photo 2-2). This particular straightedge is marketed under the name Sure-Grip® and has some unique features. It can be obtained by mail order from Trendlines (Resource 3). It comes in a variety of lengths, and the one shown is 56 in long and is designed for material up to 50 in wide. In later photos, when performing shorter crosscuts, I switch to the 30-in model for convenience. You will also need a large, plastic speed square as shown in Photo 2-3. For best results, the square should be modified by trimming approximately ¾ in off the flanges adjacent to the 90-degree corner to allow it to fit over the clamp on the straightedge as shown.

Photo 2-2

Photo 2-3

Notes:

1. To provide maximum accuracy, the workpiece must be fully supported and clamped to the plywood worktable. The saw depth of cut is then set to just barely penetrate through the workpiece and into the plywood worktable.
2. The drawings and photos in this section are shown for construction of the passenger's side triax. The driver's side is the mirror image.

2. Next, set your circular saw to cut a 45-degree bevel and make the cut shown in Photo 2-4.

Photo 2-4

3. Carefully measure the distance from the outside edge of the circular saw blade to the saw guide as shown in Photo 2-5. Add this dimension to the desired width to be cut, which is 7¾ in for the top coax adapter board as shown in Photo 2-6. Repeat this process for the remaining boards.

Photo 2-5

Photo 2-6

4. With all of the pieces cut to size, place the stack shown in Photo 2-7 in your car and verify that it fits. If you have to stagger any of the pieces to clear obstructions, match mark the pieces carefully so that you can adjust relative positions during the rest of the fabrication. For the sake of this discussion, we will assume that no offsets are required.

5. Referring to Figure 2-2, locate the triax center on each of the top four boards by drawing a horizontal line 3½ in from the top edge and a vertical line in the center. For the 9½-in-wide set, this is 4¾ in.

6. Make a photocopy of the 5⅝- by 8⅜-in oval template shown in Figure 2-3 and carefully cut it out. Align the centerlines and trace the outline on each of the four top boards as shown in Photo 2-8.

Photo 2-7

7. To locate the four mounting holes, you will need the old 6 by 9 speaker from your car. If you don't already have one, you can buy one from a junkyard for almost nothing. Place the 6- by 9-in unit facedown on the board with its major (long) axis centered and parallel with the length of the board. Carefully trace its outline and hole pattern on the board (Photo 2-9).

8. Using a compass, draw a 3⅝-in circle centered 2¼ in from one end of the board as shown in Figure 2-2. (*Note:* Be sure not to oversize the hole as the JVC coax flange is very narrow in places.) Next draw the 2⅛-in arc that defines the outer wall of the coax partition.

Photo 2-8

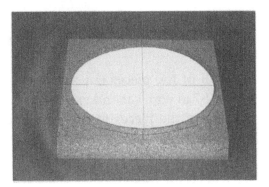

9. Cut out the hole for the coax with a saber saw as shown in Photo 2-10; then place the coax in the hole and verify proper fit. Now trace the outline of the coax flange onto the board.

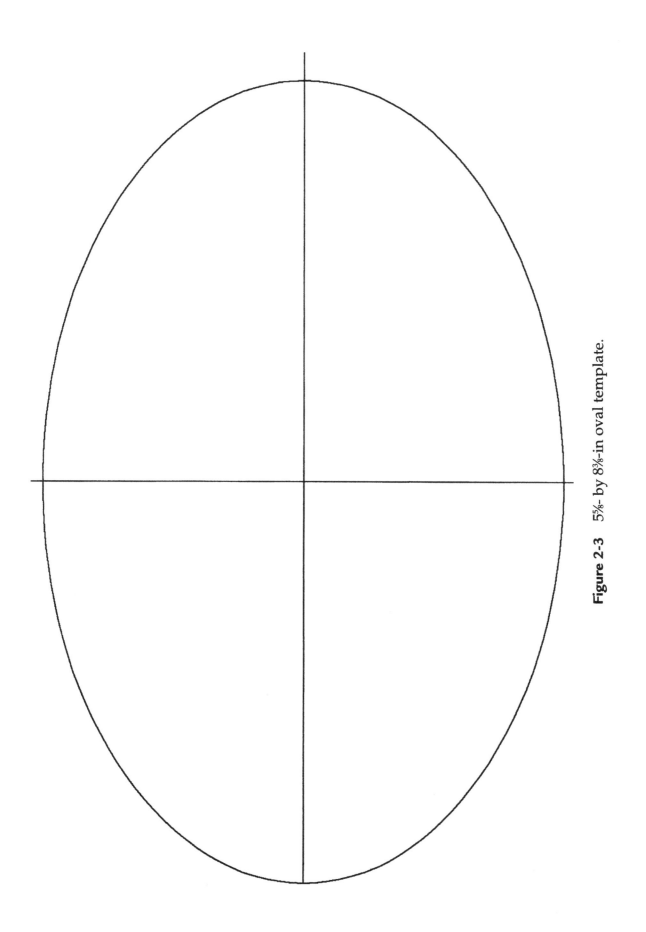

Figure 2-3 5⅝- by 8⅜-in oval template.

Photo 2-9

Photo 2-10

Align its hole pattern to avoid the 6- by 9-in hole pattern drawn earlier. Recognize that no more than two screws are actually required to fasten the coax since you will be gluing it in with silicone caulk.

10. Cut out the rest of the 5⅝- by 8⅜-in oval hole, being careful not to break the ¼-in-thick coax partition (Photo 2-11). Photo 2-12 shows the completed top coax adapter board with the

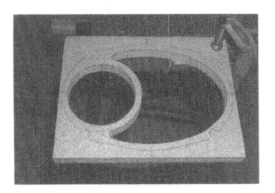

Photo 2-11

Photo 2-12

coax in place. Drill out the 6- by 9-in standard mounting holes for a slip fit with No. 8 sheet-metal screws. Then turn the board over and counterbore them just deep enough to recess the sheet-metal screw heads—about ⅛ in as shown in Photo 2-13.

11. Take the bare coax mounting board back to your car and verify that it fits and that all four mounting screws can be engaged. Incidentally, if your car has no mounting holes, now is the time to drill them. Take the board back to your workbench.

12. If you haven't already done so, solder approximately 2-ft-long leads to the coaxes and the woofer before proceeding further. Be sure to observe polarities by attaching the red lead to the terminal marked with either a + sign or a red dot.

13. Turning to the ⅜-in-thick top spacer board, cut out the 5⅝- by 8⅜-in oval opening. Place the coax facedown on this board in the same position and orientation as it will be on the adapter board. Carefully trace its outline with a pencil. Recess the marked area indicated in Figure 2-2 the minimum amount needed to fit the coax flange and provide a flat sealing surface with the coax adapter board. A router is a handy tool to use for this operation; however, a sharp router bit in an electric drill and a steady hand does a pretty fair job here also.

14. After you have confirmed fitup of the coax into the top board and the matching spacer board, clamp these two pieces together and drill the 6 by 9 mounting holes through the top spacer. Next, test-fit the two pieces in your car. If all is well at this point, take these pieces back to the workbench. Lay the first coax board on top of the second coax board and trace the cutouts for the woofer and coax. Repeat this procedure for

Photo 2-13

Photo 2-14

Photo 2-15

the third coax board. Please note the difference between the third board and the other two—it has a routed blind hole that houses the coax magnet and forms the bottom partition between the coax and the woofer pressure wave.

15. It is important that all air passages have smooth transitions, so clamp all four top pieces together and smooth off any rough edges with the rasp and sandpaper. The finished stack is shown in Photo 2-14.

16. Cut out the hole in the woofer board and fit the woofer to it. After drilling all of the pilot holes for the woofer mounting screws, fit all of the pieces together and correct any problems. Verify that the woofer surround is completely clear of the edge of the hole as shown in Photo 2-15. We are now ready to begin final assembly.

Triax Assembly and Installation

(The following procedure assumes that all pieces have been drilled and screws test-fit for final assembly.)

Photo 2-16

1. Prepare the coax for mounting by trimming off the portion of the mounting flange that protrudes past the partition and into the woofer port. Use only metal-cutting shears for this to avoid generating metal shavings. (Photo 2-16 shows a close-up of this detail.) Then, mount the coax into the top board with ¾-in-long, No. 6 panhead sheet-metal screws and silicone caulk. After this, apply a continuous bead of caulk to the top surface of the coax board, including the coax flange area, and install the spacer board on top with the ¾-in-long, No. 6 flathead sheet-metal screws.

2. Apply another bead of silicone caulk to the top of the spacer board and screw this assembly onto the bottom of the package shelf with 1¼-in-long, No. 8 panhead sheet-metal screws. Then using silicone caulk between each layer, glue and screw each succeeding board together with 1¼-in-long, panhead sheet-metal screws until all the boards are installed. Finally, mount the woofer with 1-in-long No. 8 panhead sheet-metal screws and you're done. The complete assembly is shown in Photos 2-17 and 2-18 (page 36).

Measured System Performance

As stated in Chapter 1, the frequency response of any speaker system is largely influenced by the particular installation. Vehicle cabin volume and dimensions vary widely. To remove the effect of these variables, we mounted the coax assembly in a 4- by 8-ft test wall constructed of heavily reinforced, ¾-in-thick plywood and measured its response at a distance of 12 in using the RTA as shown in Photo 2-19.

The test setup diagram is shown in Figure 2-4 (page 37). We drove the coax and woofer separately using a 50-W-per-channel amplifier.

Photo 2-17

Photo 2-18

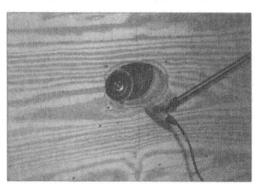

Photo 2-19

Subwoofer responses were measured with the filter described in Chapter 6, which also serves as an adjustable crossover.

Frequency responses for the composite system with the filter are 30 to 16K Hz ± 3 dB as shown in the RTA printout in Figure 2-5*a* while Figure 2-5*b* and *c* show the separate responses for the woofer and coax. Due to the shelved bass response of the coaxes, it was necessary to set the subwoofer crossover point at approximately 500 Hz to obtain a flat response in the midbass region. The sub-

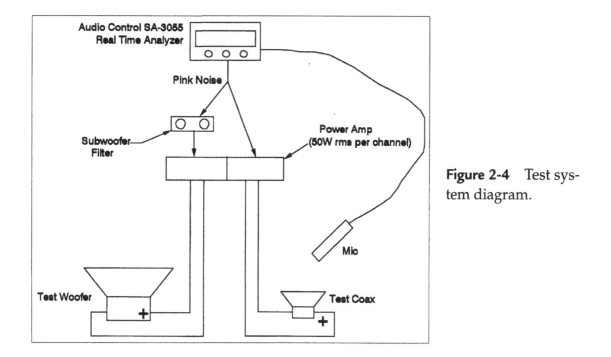

Figure 2-4   Test system diagram.

woofer filter was set for a high-pass Q of 1.7 and a low-frequency cutoff of 25 Hz.

As a point of interest, Figure 2-6a shows the theoretical woofer response with the crossover point set for a more typical 100 Hz. Figure 2-6b is the actual response of the woofer measured with the RTA with the same settings. The two appear to be very similar.

Sensitivity of the composite triax at approximately 3 V (as close as I could get to the standard 2.83) input at 1 kHz is 90 dB as measured with the RTA.

To see if the extensive baffling caused woofer distortion, I measured the output waveform of the triax at a sound pressure level of 100 dB at some key frequencies—30, 40, 500, 1K, and 10K Hz. I did this by feeding a sine wave to the system with a signal generator in place of the RTA pink noise and capturing the on-axis response with a microphone connected to a dual trace oscilloscope. The photographed oscilloscope displays are shown as Photo 2-20a through e. Each screen shows the signal generator waveform on bottom with the measured waveform above it. Photo 2-20a shows the waveform

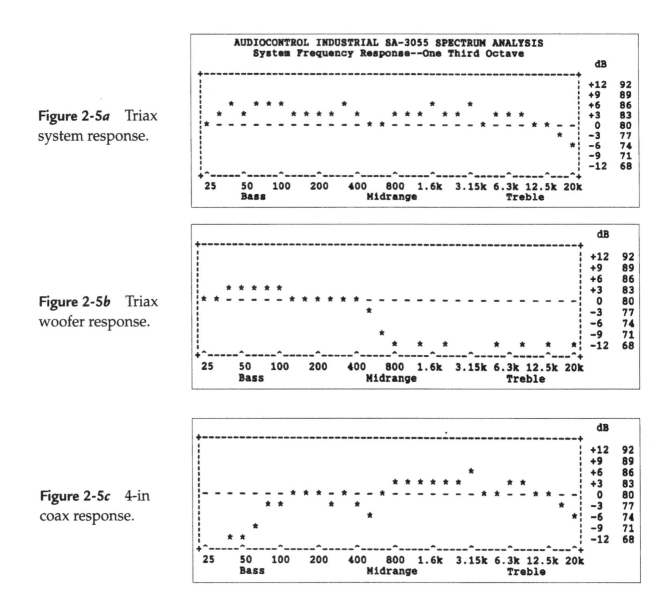

Figure 2-5a Triax system response.

Figure 2-5b Triax woofer response.

Figure 2-5c 4-in coax response.

at 30 Hz to be slightly asymmetric, indicating that the woofer is producing some distortion at 100 dB that is significantly reduced at 40 Hz in Photo 2-20b. Higher frequencies all appear to be sinusoidal.

Finally

For a purely subjective analysis, I connected a portable CD player to the test setup and played music through it. It sounded smooth and full. I then plumped up the bass by increasing the woofer volume and reduced the woofer crossover frequency to reflatten the midbass. The triax sounded surprisingly musical. As further proof, I had

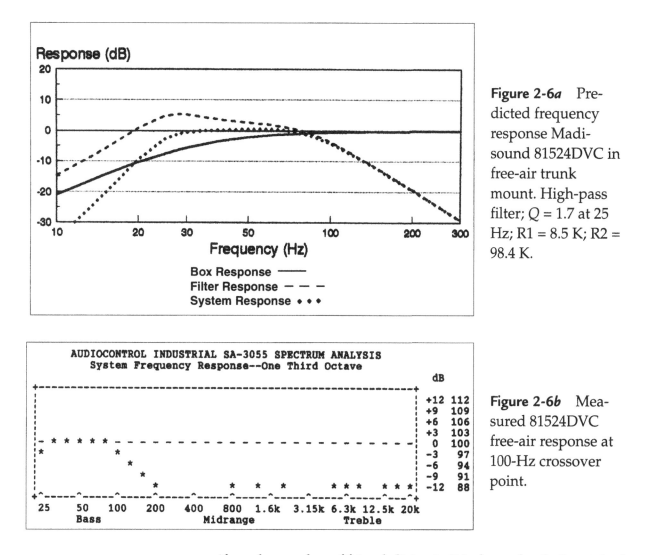

Figure 2-6a Predicted frequency response Madisound 81524DVC in free-air trunk mount. High-pass filter; $Q = 1.7$ at 25 Hz; R1 = 8.5 K; R2 = 98.4 K.

Figure 2-6b Measured 81524DVC free-air response at 100-Hz crossover point.

my wife and a number of friends listen to it independently. I watched for their immediate reactions as the music began to play. In every case, they were surprised at the quality and fullness of the sound emanating from the 6 by 9 oval hole.

I tested the power-handling capacity by increasing the volume for brief periods and recorded the results. At a distance of 12 in in free air, I measured continuous outputs of 110 dB (and more) at 40 Hz and above. In the confined space of a sedan, I estimate that a pair should be able to generate SPLs in the range of 115 to 118 dB with a 50-W-per-channel amplifier. Even higher SPLs can be obtained by using woofers with longer travel.

Photo 2-20a
(30 Hz)

Photo 2-20b
(40 Hz)

Photo 2-20c
(500 Hz)

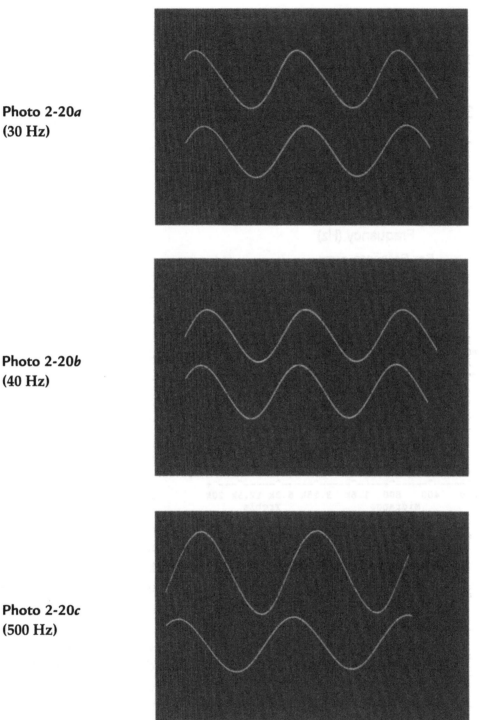

I have used this triax in several installations, one of which was my own personal vehicle. It requires quite a bit of patience and physical labor to install it properly, but the results are well worth the effort.

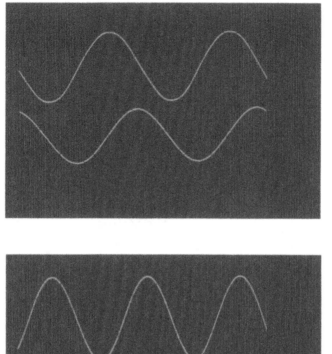

Photo 2-20*d*
(1K Hz)

Photo 2-20*e*
(10K Hz)

Bandpass/Coax Combination

Like the previous system, Sedan System No. 2 is intended to fit a 6- by 9-in oval package shelf cutout. It also combines a coax with a sub-woofer, but instead of a free-air woofer, this design employs a relatively small but powerful bandpass box. Figure 2-7 shows the general arrangement. A simple adapter plate couples a 5-in coax and the bandpass box's 3-in-diameter PVC port to the 6- by 9-in opening. In most respects, construction and installation are simpler than for the triax. The main trade-off is that the boxes take up more trunk room.

Photo 2-21 shows the two Madisound drivers selected for this project. To minimize box volume, I selected the 8154 8-in woofer. Because more room is available on the adapter plate, we will switch to a 5-in 5402 coax, which employs the Audax 0.5-in polymer

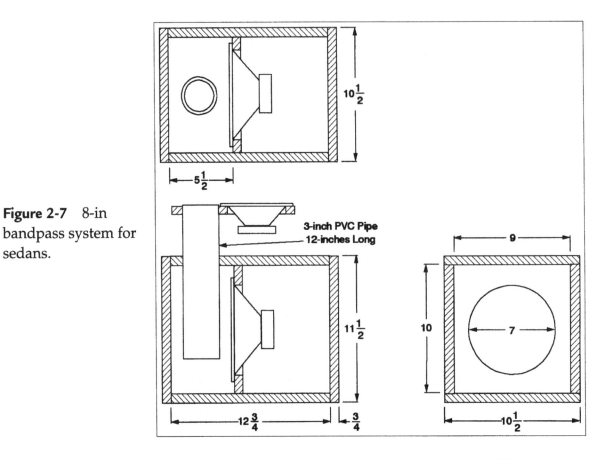

Figure 2-7 8-in bandpass system for sedans.

tweeter. This coax has much higher power handling and more extended frequency response than the 4-in JVC. It will also provide more midbass punch and is a better match for the bandpass box. For a sneak preview, Figure 2-10*a* (page 55) shows the measured response of the complete system.

I designed the bandpass box using some shareware—*Loudspeaker CAD Lite,* which I downloaded off the Internet. I'll discuss this program and a few others like it in more detail in Chapter 8. The predicted response is shown in Figure 2-11*c.* The characteristic efficiency for 1-W input is predicted to be a reasonable 88 dB at a distance of 1 m, which is dependent on the particular combination of box volumes and tuning frequency chosen. This design can be employed with or without the electronic subwoofer filter, but it sounds best with the filter since the filter can be set to increase bass extension and remove subsonic information. It can also be used to adjust the high-frequency roll-off to better match the coax. Actual

Photo 2-21

response curves with and without the filter are shown later on in the chapter.

The tools required for this project are the following:

- Circular saw
- Saber saw
- Variable-speed drill with drill bits and Phillips screwdriver bit
- Wood rasp
- Router (optional)
- Belt sander (optional)
- Straightedge
- Square
- Worktable
- Ruler, tape measure
- Clamps (optional)

The materials required are the following:

- Two pieces of ¾-in-thick particle board shelving material, 11½ in wide by 8 ft long
- Box of 100 No. 6 coarse-threaded drywall screws—1⅝ in long
- 16—No. 8 panhead sheet-metal screws—1 in long
- 8—No. 6 panhead sheet-metal screws—¾ in long
- Yellow carpenter's glue
- An 8-in length of 3-in Schedule 40 PVC pipe
- Rope caulk
- Silicone caulk
- Madisound 8154 8-in woofer (one or two as desired)

- Madisound 5402 coaxes (pair)
- Terminal cup (one or two as desired)

Construction of the project is as follows:

1. Referring to Figure 2-8, begin by ripping all of the pieces to width using the setup shown in Photo 2-2. Or, if you're fortunate enough to have access to a table saw, this is a good application for it.

   Note: Making the top and bottom slightly wider and the end pieces slightly larger than plan will eliminate the possibility of creating unsightly underlapped edges and permit you to trim off the overlaps flush with a router or belt sander after final assembly. Skip ahead to Photo 2-30 (page 48) to see how this should look. I typically make the overlaps ¹⁄₁₆ in.

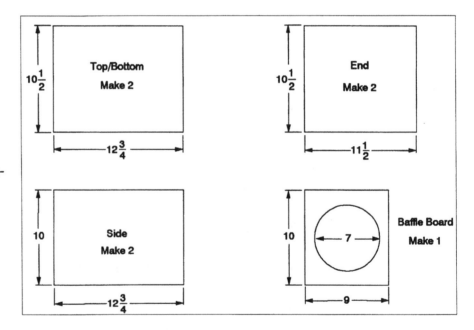

Figure 2-8 Bandpass box cutting diagram.

2. Photo 2-22 shows the setup for cutting the sides and top to the same length. You must first install a stop block on one end of the workbench that is taller than ¾-in thickness of the material. This block must be securely fastened and be as perpendicular to the workbench surface as possible.

3. Next, cut a piece of scrap to a length equal to the width of the saw plus the length of the side and top pieces—12¾ in. Use this

Photo 2-22

measuring tool to set the length of cut for each of the four pieces as shown. The first cut is shown in Photo 2-23. Again, set the depth of cut to barely penetrate the workbench surface.

4. Cut the hole for the 8-in woofer in the baffle board as shown in Photo 2-24. Test-fit the driver as shown in Photo 2-25. Carefully mark the location of the mounting holes. After removing the driver, center-punch and drill the pilot holes using a $\frac{3}{32}$-in drill. We will be mounting the driver with 1-in-long, No. 8 panhead sheet-metal screws. To facilitate assembly later, go ahead and run a screw through each of these holes.

Photo 2-23

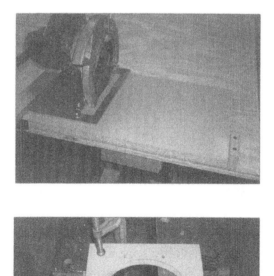

Photo 2-24

Photo 2-25

5. Next, cut the 3½-in hole in the top for the 3-in PVC pipe. For best results, be careful to avoid oversizing this hole. As a rule, I prefer to initially cut portholes slightly undersized and then carefully enlarge them with a wood rasp for a snug fit.
6. A 3-in-diameter hole is needed to mount the terminal cup as shown in Photo 2-26. After verifying fitup, carefully mark and drill the four mounting holes for some ¾-in-long No. 6 pan-head sheet-metal screws using a ¹⁄₁₆- or ⁵⁄₆₄-in drill.
7. Carefully lay out all of the screw holes on the outer surfaces of the rest of the pieces. Some general guidelines should be followed.

Photo 2-26

- Screw holes should be between 2 and 3 in apart and be evenly spaced.
- For ¾-in material, the holes should be ⅜ in from the edges.
- To minimize splitting the particle board, avoid locating holes less than 1¼ in from the ends.
- After all the holes are located, center-punch and drill with a ⁵⁄₃₂-in drill for a slip fit. The best and easiest method for countersinking the holes evenly is to install a secure stop

on a ⅜-in drill bit as shown in Photo 2-27. Without a stop, it is difficult to avoid occasionally having either a drill or countersink bit grab, which results in excessive countersink depth. I have only been able to find the stops through mail order catalogs such as Resources 3 and 4.

Photo 2-27

8. Carefully draw lines on the inside surfaces of each of the pieces to represent the location of the mating parts. Drive some small nails along these lines to act as stops to accurately locate the pieces during final assembly as shown in Photos 2-28 and 2-29. Note also the use of the corner clamp to hold the pieces in place. A couple of these are better than an extra pair of hands.

9. As you carefully dry-fit each piece, drill the pilot holes in the edges of the mating parts by running a ³⁄₃₂-in drill through the layout holes. We will assemble the box with the drywall screws.

10. The dry-fit, assembled box is shown in Photo 2-30. Note the slight overlap on the top, bottom, and ends.

11. It's now time to make the port. Start by cutting a length of 3-in PVC pipe to a length of 8 in. While this is typically done with a hacksaw, it is much easier to do on a table saw using

Photo 2-28

Photo 2-29

Photo 2-30

the miter gage, as shown in Photo 2-31. The inner and outer edges of the pipe must be rounded and smooth to minimize port noises. This can be done with a suitable file and sandpaper, but it is easily done on a router table with a ¼-in-radius roundover bit as shown in Photo 2-32.

12. With all the prep work done, we're ready for final assembly. I recommend the following procedure:

- Place a bucket of warm, clean water and a suitable clean rag near the workbench to wipe off excess glue.

Photo 2-31

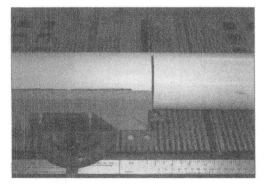

Photo 2-32

- Temporarily install the right side to the bottom and glue and screw the baffle board to the bottom. Use *plenty of yellow carpenter's glue* and wipe off the excess with the damp rag, which you should rinse often. Install only two or three screws in a piece at this point in case you have to remove it to make an adjustment.
- Remove the right side and reinstall with glue and screws.
- Glue and screw the rear panel in place.
- Glue and screw the left side in place.
- Dry-fit the front panel in place and install several screws.
- At this point you should be ready to install the top as shown in Photo 2-33. Note the copious amount of glue applied.

Photo 2-33

- After the top is screwed into place, note the amount of glue that squeezes out in Photo 2-34. This is typical for a properly done joint.
- After wiping away all the excess glue, *remove the front panel,* and allow the box to dry overnight.
- Reinstall the front panel *dry* with a minimum of eight screws, but be sure to drill all of the pilot holes at this

Photo 2-34

point. Now trim off the overlaps (if you chose to make any) with a router using a flush trim bit as shown in Photo 2-35. Next round over the edges with a ⅜-in-radius roundover bit as shown in Photo 2-36. I chose an inexpensive bit for this operation in case it happens to hit a screw head.

Photo 2-35

- Take the box back to the workbench and again remove the front panel.
- In Photo 2-37, the woofer is prepared for mounting by soldering on about 14 in of speaker wire and placing two rows of rope caulk on the back of the mounting flange.
- In Photo 2-38, the woofer is shown screwed into place and a bead of silicone caulk is being applied in preparation for final installation of the front panel.
- The terminal cup connections are made and soldered as shown in Photo 2-39. Now screw the cup into place. To avoid distorting the cup and creating a leak, do not over-tighten.
- In Photo 2-40, the pipe is positioned in the hole and a ring

Photo 2-36

Photo 2-37

of silicone caulk is applied. The pipe is inserted in a twisting motion that evenly distributes the caulk as shown in Photo 2-41. The depth of insertion can be adjusted as necessary to fit your particular vehicle. However, a minimum depth of approximately 2 in is recommended to avoid affecting system tuning. Remove the excess caulk with a putty knife and allow to cure overnight.

13. We're now ready to begin construction of the adapter board. Begin by cutting a piece of ⅝-in-thick particle board to the

Photo 2-38

Photo 2-39

Photo 2-40

dimensions shown in Figure 2-9. Photo 2-42 shows the partially complete board. Photo 2-43 shows the coax being mounted while Photo 2-44 shows the coax and a scrap piece of pipe in place to verify fitup. With the adapter board complete, we're ready for testing.

Free-Air Testing

We mounted the entire assembly in the test wall as shown in Photo 2-45 and connected the standard test setup of Figure 2-4. Fig-

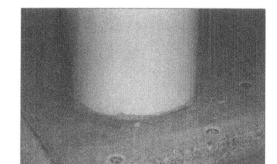

Photo 2-41

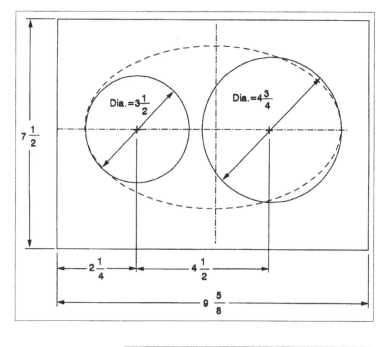

Figure 2-9 Band-pass/coax adapter board.

Photo 2-42

ure 2-10*a* to *c* show the resulting system response with the filter set for a 6-dB boost at 30 Hz (R1 = 9K, R2 = 63K). The composite frequency response is approximately 35 to 20 kHz ± 3 dB. Figure 2-10*b* and *c* show the subwoofer and coax responses, respectively.

Photo 2-43

Photo 2-44

Photo 2-45

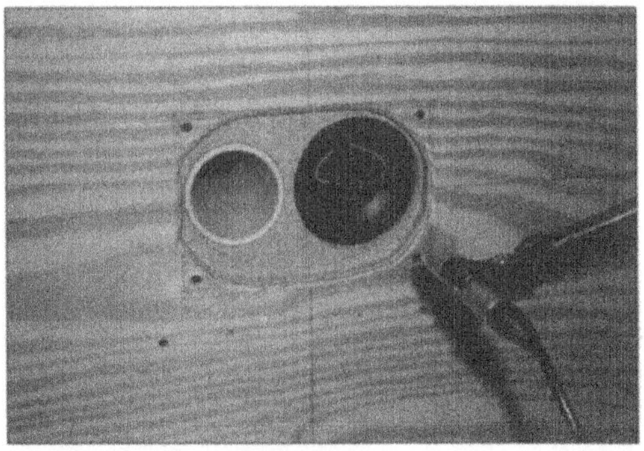

Photo 2-46*a* to *f* show output waveforms at 35, 40, 50, 500, 5,000, and 10,000 Hz. At 35 Hz, the waveform is a little asymmetric but it is essentially cleared up at 50 Hz. All of the rest of the waveforms appear sinusoidal.

Subjectively, the sound of this system is punchy and authoritative. Because of the robust coax, midbass is solid and drum kicks have good leading edge attack. Bass is solid, but port noises can be heard above 103 dB in free air. The Audax tweeter is crisp and clear,

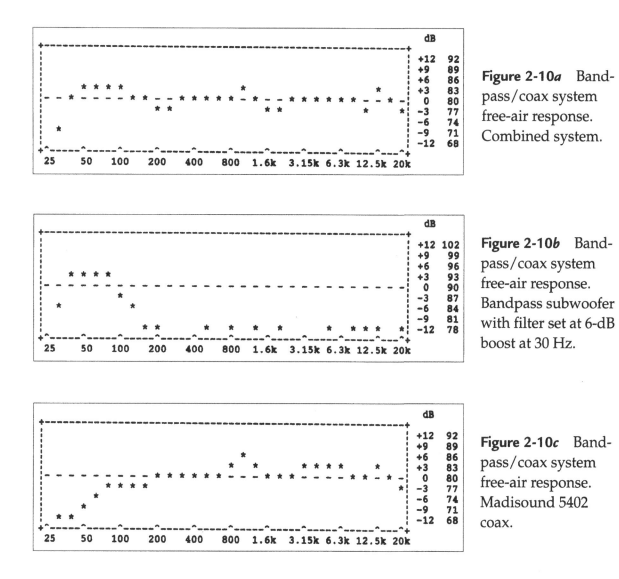

Figure 2-10a Bandpass/coax system free-air response. Combined system.

Figure 2-10b Bandpass/coax system free-air response. Bandpass subwoofer with filter set at 6-dB boost at 30 Hz.

Figure 2-10c Bandpass/coax system free-air response. Madisound 5402 coax.

but a little edgy. In any case, the coax will be toned down considerably when used in the rear fill mode, and should be excellent in that application.

In-Car Testing

In-car testing showed that another plus for this system is high sensitivity. Under these conditions, a sound pressure level of 96 dB was recorded for an input of 2.8 V rms, which amounted to a paltry 1.5 W! Maximum in-car SPL achieved with a single unit was 109 dB before port noise became intrusive, so a pair should be able to generate 112 to 115 dB.

Photo 2-46*a*
(35 Hz)

Photo 2-46*b*
(40 Hz)

Photo 2-46*c*
(50 Hz)

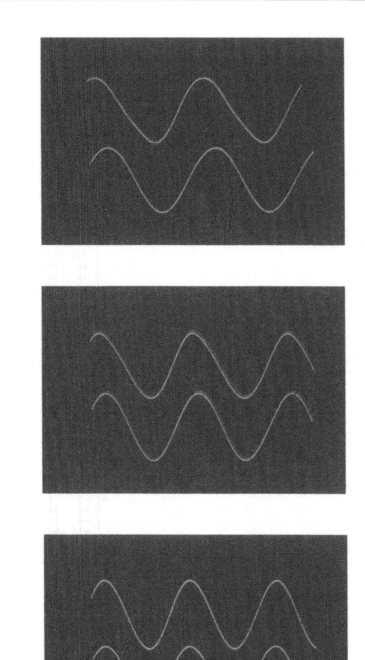

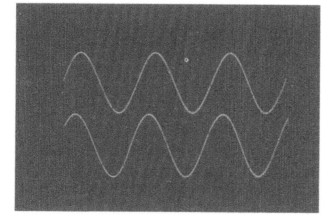

Photo 2-46*d*
(500 Hz)

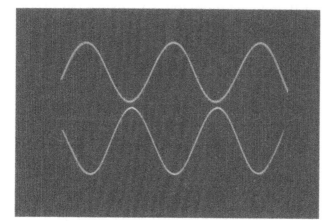

Photo 2-46*e*
(5000 Hz) or
(5K Hz)

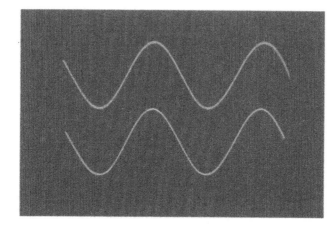

Photo 2-46*f*
(10,000 Hz) or
(10K Hz)

Figures 2-11*a* and *b* show the subwoofer in-car response with and without the filter. Without the filter the box has a peak at about 90 Hz and shelves down 3 dB out to 40 Hz. This conforms closely to the predicted response shown as Figure 2-11*c*, which clearly shows the slope in the passband with the port purposely shortened from the target 10 to 8 in to reduce port noise and overhang. With the filter, the response is flat from 40 to 80 Hz, except for the dip at 70. The sound of this system is clearly much more pleasing with the filter. All higher-frequency content becomes inaudible and the boost at 30 Hz followed by the rapid cut eliminates the overhang that can be heard without the filter. It simply sounds much tighter. On the other hand, if you don't plan to use the filter, you may want to try a port length of 10 in to get more low-bass extension. When this is used in combination with the low-pass filter built into many amplifiers, the sound may be acceptable, although port noise may be more troublesome.

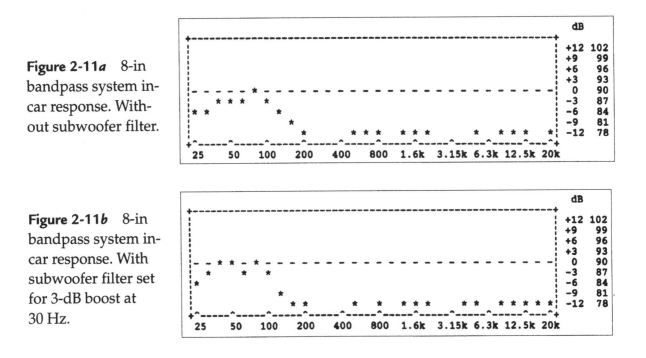

Figure 2-11*a* 8-in bandpass system in-car response. Without subwoofer filter.

Figure 2-11*b* 8-in bandpass system in-car response. With subwoofer filter set for 3-dB boost at 30 Hz.

Final Comments

Overall, the bandpass/coax rear speaker system sounds very good. While the subwoofer appears to have less headroom than some of the other systems you will encounter in this book, a pair of them should provide adequate output for most types of music. Bass-heavy music like rap may create excessive port noise at high volumes. The boxes take up more trunk space than I would prefer but are still quite compact compared to other systems I have seen stuffed into the trunks of some car stereo enthusiasts. Because they are remarkably efficient, amplifier power is not much of a consideration. Installation is relatively easy and fabrication difficulty is about the same as that for any box system. And with a little ingenuity, it can easily be adapted to fit virtually any sedan—not just those with 6 by 9 oval speaker cutouts.

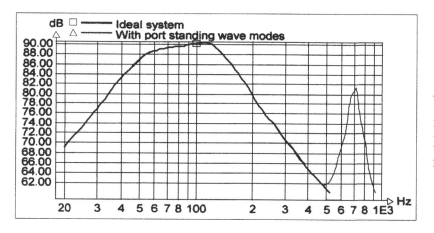

Figure 2-11c 8-in bandpass system in-car response. Predicted free-air response.

Bass Couplers

The two preceding systems are compromises intended to minimize or eliminate the need to do any modifications to your car's interior. If you remove that restriction, a more direct route to better sound may be found in one of the three alternative systems shown in Figure 2-12. By moving the rear satellite speakers outboard of the factory 6 by 9 cutouts, we can free up these relatively large openings for bass reproduction. The advantage to this approach is that you can get the best of both worlds—punchy midbass and extended headroom. The suggested method is to combine a pair of 5- or 6-in coaxes with two or more 8-in woofers by using one of the bass couplers shown in Figure 2-13 or 2-14.

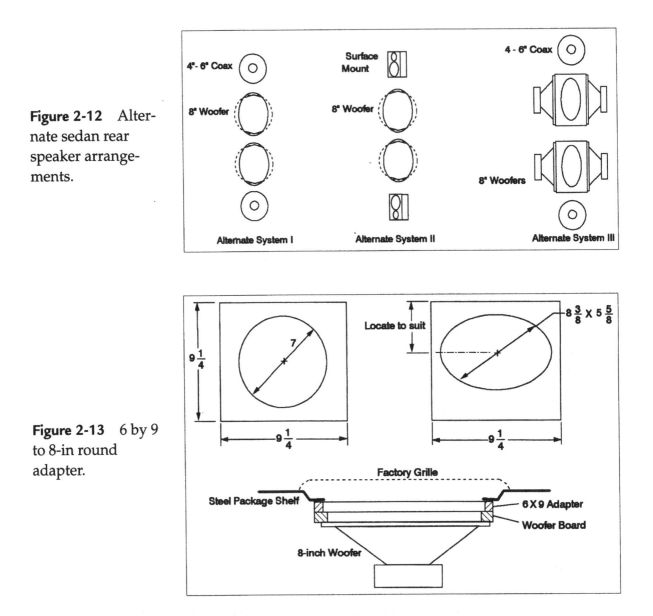

Figure 2-12 Alternate sedan rear speaker arrangements.

Figure 2-13 6 by 9 to 8-in round adapter.

There is one drawback to this scheme. You will have to alter the rear package shelf. While some would not find this a problem, others would be concerned about possibly devaluing their expensive vehicle (and rightfully so). In either case, cutting into the steel package shelf is probably best left to a professional installer. If done properly, the work will be well worth the fee incurred.

Alternate System I employs two 8-in woofers firing through the adapter boards shown in Figure 2-13. This is a very good sounding system and easy to fabricate and install. You can make these adapter

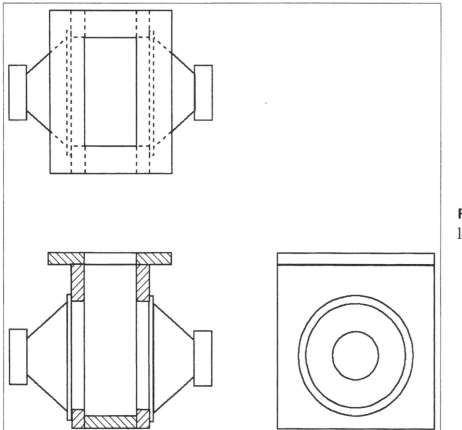

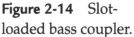

Figure 2-14 Slot-loaded bass coupler.

boards with literally nothing more than a drill and a saber saw, and installation is equally easy.

Alternate System II is similar to System I except here the coaxes are replaced by surface-mounted two-way systems. The sound of this system becomes largely dependent on the quality of the surface mounts—which can vary widely. These need to be purchased only after careful auditioning. On the positive side, package shelf modifications will be minimal, consisting of a few drilled holes to attach the speakers and run the wiring. On the negative side, you may not like the look.

Alternate System III is considerably more powerful than I and II. It employs two *pairs* of slot-loaded 8-in woofers using the coupler shown in Figure 2-14. The combined woofer cone area is somewhere between two 10-in and two 12-in woofers—all firing efficiently through a pair of 6- by 9-in openings!

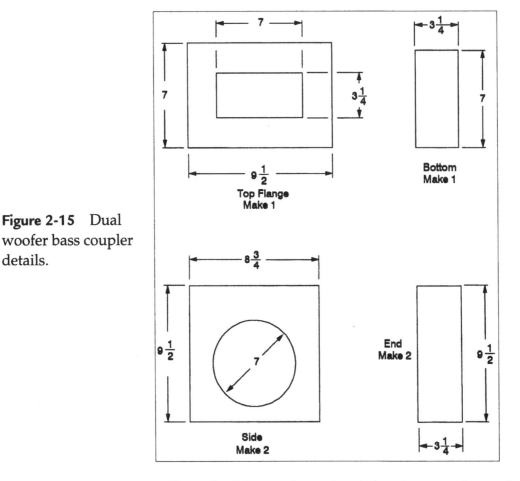

Figure 2-15 Dual woofer bass coupler details.

To make this coupler, cut out the pieces as shown in Figure 2-15. Then lay out, drill, and countersink all of the screw holes. Note that the inside edge of the woofer holes should be rounded over with a router or file to reduce turbulence. I used a ¼-in roundover bit, but a ⅜-in radius might be even better. The objective is to roundover as much as possible without weakening the screw holes for mounting the drivers.

Photo 2-47 shows all the pieces cut out and the assembly process begun. Here one of the baffle boards is shown clamped in position to be dry-fit to one of the side pieces. Photo 2-48 shows the partially assembled coupler while Photo 2-49 shows the completely assembled coupler with the woofer hole edges filled with water-based wood putty and ready for sanding. Note the two 45-degree blocks placed in the inside corners to aid in launching the pressure wave toward the slot. Also, I purposely made the slot a little smaller than the final opening so that I could come back with the router and

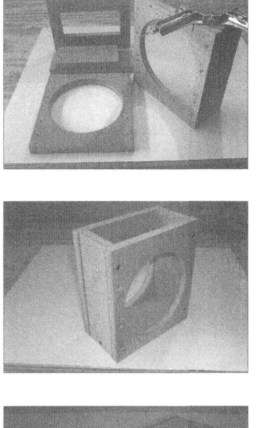

Photo 2-47

Photo 2-48

Photo 2-49

trim the edge flush with the walls as shown in Photo 2-50. The assembled unit is shown in Photo 2-51 using two Madisound 81524-DVC woofers. As was done for the triax, only one voice coil per driver is connected to obtain the proper Q_{TS} for this free-air application.

I mounted the coupler in the test wall and connected the test rig with the subwoofer filter shown previously. After some experimentation, I obtained the response shown as Figure 2-16 by setting the

Photo 2-50

Photo 2-51

filter for a modest 2-dB boost at 30 Hz (R1 = 11.4K, R2 = 58.3K). The response shown is flat from 30 to 80 Hz and is down 3 dB at about 28 Hz. The crossover frequency was arbitrarily set at 100 Hz. The woofers reached their full cone excursion limits by applying a mere 1.5 V rms input at 35 Hz. Sound pressure levels of 105 dB were recorded at a distance of 12 in in free air. At higher frequencies, the system easily produced 110 and 115 SPLs at 1 m, which agreed well with the value of 112 dB predicted by Perfect 4.5—a shareware speaker design program.

Figure 2-16 Slot loaded bass coupler free-air response with subwoofer filter.

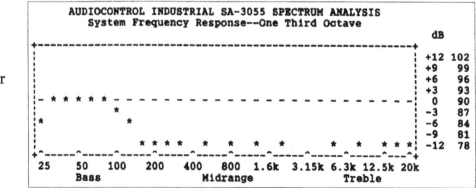

Results of sine-wave testing at a 100-dB SPL are shown in Photo 2-52 for four low-frequency points. Some small amount of asymmetry is seen at 35 Hz, which clears up by 40 Hz. At 70 and 100 Hz, waveforms are clean and sinusoidal but the phase shift due to the filter comes into play. Perhaps a better way to display these particular waveforms would be without the filter. Waveforms at all higher frequencies were quite clean, indicating that the preparations done to the hole edges appeared to be beneficial.

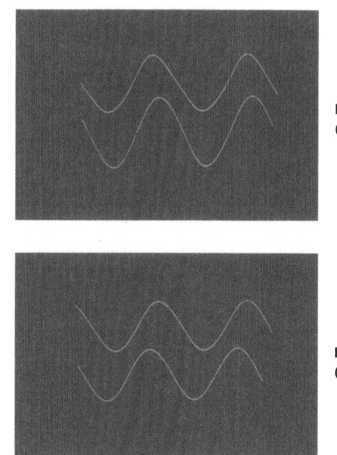

Photo 2-52*a*
(35 Hz)

Photo 2-52*b*
(40 Hz)

Subjectively, the sound of this system is extremely smooth and effortless. The sound pressures obtainable are limited only by the cone excursion limits of the drivers selected. With a single pair of Madisound 81524-DVCs, which have a peak-to-peak excursion limit of 5 mm, in-car SPLs should easily reach 115 dB—and all this with a handful of watts! Two pairs should generate in-car SPLs of 120 dB. If you need even more, you can install drivers with longer travel.

**Photo 2-52c
(70 Hz)**

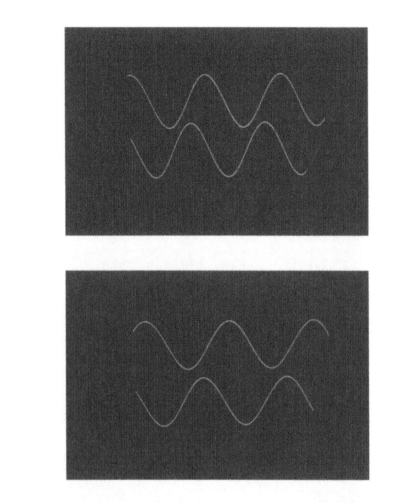

**Photo 2-52d
(100 Hz)**

That wraps up this chapter on sedan rear speaker systems. I have attempted to provide some unique and (I hope) useful solutions to the difficult sedan bass problem that won't break your bankroll or damage your car. Pick the one that best suits your needs and construction capabilities.

Chapter 3

Rear Speaker Systems for Hatchbacks

Rear speaker systems for hatchback cars tend to be more simple and straightforward than for other types of vehicle. There is usually quite a bit of space available and we can usually get by with a rectangular box. Size is limited only by the amount of cabin volume you are willing to give up. Figure 3-1 shows two possible speaker scenarios for a hatchback system. System I is based on the assumption that the vehicle is

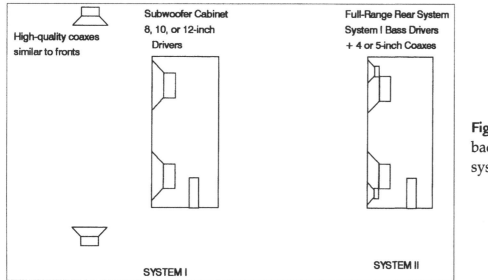

Figure 3-1 Hatchback rear speaker systems.

equipped with rear speaker receptacles somewhere in the cabin walls. In this case, we would replace the factory speakers with some quality coaxes and add one of the subwoofers we will build in this chapter. On the other hand, if your car was not provided with rear speaker receptacles, System II is a full-range unit which has them built in. It contains both a pair of subwoofers and a pair of high-quality coaxes in their own sealed chambers. We will build and test one version of this system also.

Getting great bass from a subwoofer in a hatchback is probably easier than in any other type of vehicle. The speaker placements shown in Figure 3-2 make it easy to see why. The tapered cabin space

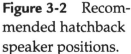

Figure 3-2 Recommended hatchback speaker positions.

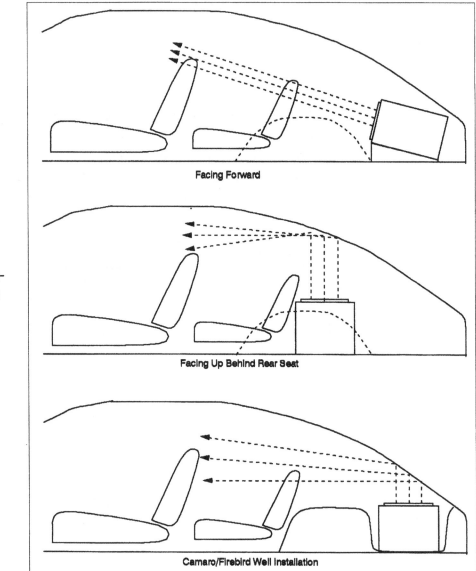

acts like a large horn, which acoustically amplifies the output from the subwoofer. The system is so efficiently coupled to the cabin air space that minimal amplifier power is needed to generate fairly high sound pressure levels and not much of this escapes outside the vehicle. More than one client has raved about these attributes, which are somewhat amazing, as the test data will bear out.

System 1

We begin our construction with what I consider to be one of the finest-sounding subwoofers I have ever encountered. I have built this system in every format imaginable and have never been disappointed with the results. Its main attributes are small size—1 ft^3 net volume for two 8-in woofers, extended low-frequency response—3 dB down at 33 Hz in free air, and high-power handling—up to 200 W of music power. This particular box is sized to fit into the well of 1990 and up Camaros and Firebirds but can be reconfigured to fit many applications as long as interior volume and port dimensions are maintained.

The tools required are the same as used on the projects in Chapter 2. All of the straight cuts can be made with a circular saw and clamp-on straightedge. A saber saw is needed to cut the holes for the drivers and terminal cups and port. Since we will be applying fiberglass insulation to the interior walls, we'll need a medium-duty staple gun.

The materials required are the following:

- ¾-in-thick particle board—two 8-ft lengths of 11½-in-wide shelving
- Box of coarse-thread drywall screws, 1⅝ in long
- Yellow carpenter's glue
- Box of rope caulk
- 8 No. 6 panhead sheet-metal screws, ¾ in long
- 1 ft of 3-in Schedule 40 PVC pipe
- 1 package of fiberglass insulation from Radio Shack
- 2 Madisound 8154 8-in woofers with grilles
- 2 terminal cups
- 3 ft of No. 16 or 18 speaker wire
- Solder

The Madisound 8154 drivers used in this project are relatively inexpensive and are shown in Photo 3-1 along with the terminal cups, the only other parts required.

Photo 3-1

Figure 3-3 shows the general arrangement and cabinet dimensions. If you have a router, make the front, back, and end pieces oversized so you can trim them off flush after final assembly.

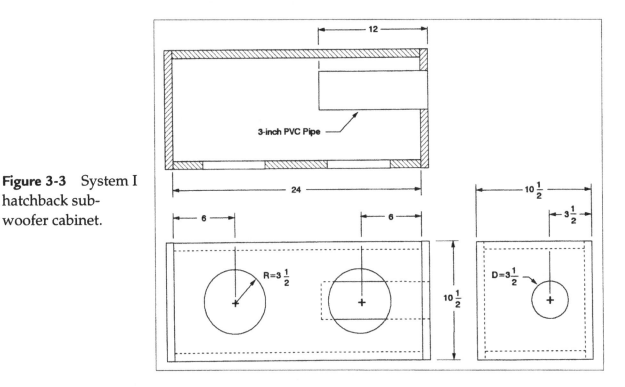

Figure 3-3 System I hatchback sub-woofer cabinet.

Photo 3-2 shows all the pieces cut out and ready for assembly. As always, we must first dry-fit the pieces together to verify that they are correctly sized. We begin by attaching the back to the top as

Photo 3-2

shown in Photo 3-3. Here, using a corner clamp at the top makes this much easier. Next, attach the bottom to the back, followed by the front. Lastly, square everything up and attach the ends. The dry-assembled box is shown in Photo 3-4. In Photo 3-5, we can see the generous amount of excess glue being squeezed out as the top is

Photo 3-3

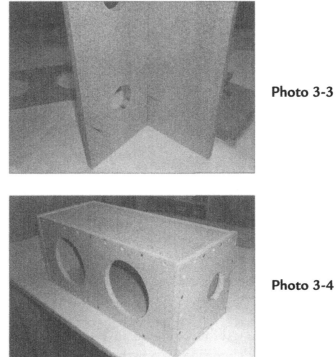

Photo 3-4

glued into place. (Don't forget to have the wet towel at the ready.) This will ensure both good bonding and an airtight seal. Photo 3-6 shows the completely assembled box. Note the overlaps on the ends and top pieces. We trimmed those off flush with the router as shown in Photo 2-35. The result is shown in Photo 3-7.

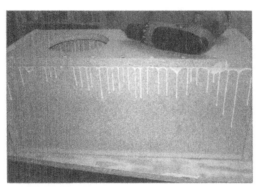

Photo 3-5

Photo 3-6

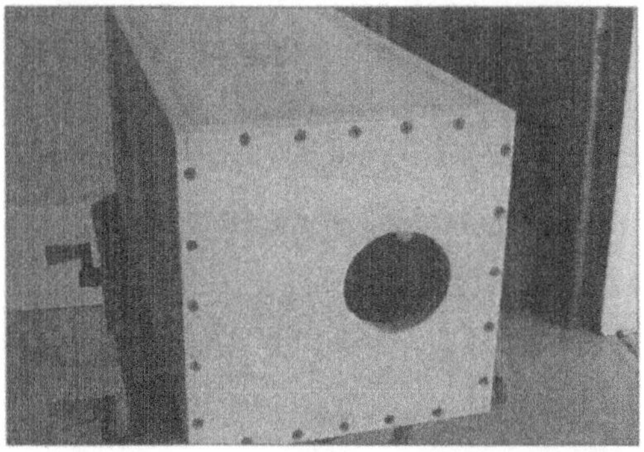

Photo 3-7

At this point we have several options as to how we treat the edges, depending on whether or not the cabinet will be painted or carpeted. If we're painting, the preferred edge treatment would be to round over with the router, fill with water-based putty, and sand smooth. If we're planning to apply carpet, we can skip the edge filling. Later on in Chapter 6 we'll use this cabinet to demonstrate proper carpeting techniques, so we'll simply sand everything smooth and round off the corners and sharp edges.

We're now ready to turn this box into a speaker system. Ported systems tend to sound smoother if they are lined with fiberglass, and an excellent grade of this material is available at your local Radio Shack. In Photo 3-8, the fiberglass has been cut to size and stapled into place. All sides but the front and the end where the port is located should be neatly covered.

Photo 3-8

In Photo 3-9, the wiring is soldered to the terminal cups and the cups are screwed into place in the back of the cabinet as shown in Photo 3-10.

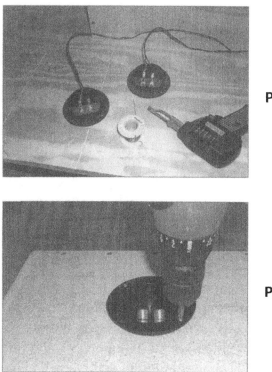

Photo 3-9

Photo 3-10

In Photo 3-11, we begin to prepare the woofers and grilles by applying two rows of rope caulk to the back of the grille ring. Photo 3-12 shows the caulk applied to the grille ring and the back of the woofer flange. The woofer is then placed into the ring, and wiring is connected and soldered as shown in Photo 3-13. Next, all eight

Photo 3-11

Photo 3-12

Photo 3-13

mounting screws are inserted through the woofer, grille ring, and the two layers of caulk (Photo 3-14). The woofer is placed into its hole and each of the screws is started by hand into the predrilled holes. The screws are tightened evenly until everything is pulled up tight as shown in Photo 3-15, which causes the caulk to be extruded out from under the grille ring as shown in Photo 3-16.

Photo 3-14

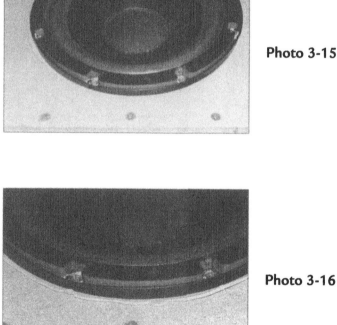

Photo 3-15

Photo 3-16

In Photo 3-17, the port is inserted into its hole and a continuous ring of silicone caulk is applied about an inch from the end of the pipe. The pipe is inserted with a twisting motion to evenly distribute the caulk as shown in Photo 3-18. I chose to leave the pipe protruding about ¼ in to prepare for the carpet, which will be applied later. The completed system without the grilles is shown as Photo 3-19.

Photo 3-17

Photo 3-18

Photo 3-19

With the grilles snapped into place, the system is ready for testing in the well of a 1995 Pontiac Firehawk as shown in Photos 3-20 and 3-21. (This particular Firehawk has T-tops which are stored in the well when removed. So this well has some fittings in the bottom which hold the box about 2 in off the bottom.)

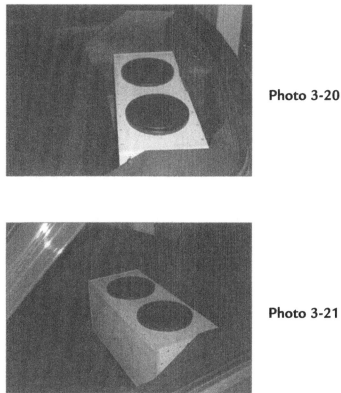

Photo 3-20

Photo 3-21

With the hatch closed and all the windows up, I performed some preliminary pink noise testing using the standard test setup. The system had plenty of low-end extension, so I set the filter for flat response with a cutoff at 30 Hz for rumble suppression. The result is shown in Figure 3-4*a* and *b*. In Figure 3-4*a*, I set the drive level for

```
                                                                  dB
+------------------------------------------------------------+
|                                                            | +12 102
|                                                            | +9   99
|                                                            | +6   96
|                                                            | +3   93
|- * * * * * - - - - - - - - - - - - - - - - - - - - - - - - | 0    90
|            *                                               | -3   87
|                                                            | -6   84
|*             *                                            | -9   81
|                * *     *   *   *   *     *   *   *   *    | -12  78
+^-----^-----^-----^-----^-----^-----^-----^-----^-----^---^+
 25    50   100   200   400   800  1.6k  3.15k 6.3k 12.5k 20k
       Bass                   Midrange              Treble
```

Subwoofer Filter set for 0 dB Boost with 30 Hz cutoff (R1=17.1K, R2=34.1K).
Drive level = 0.7 volts rms/channel (approx. 0.1 watt/channel)

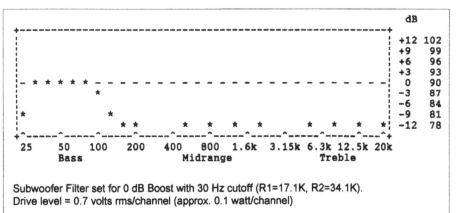

Figure 3-4*a* Hatchback System I in-car frequency response at 90 dB.

90-dB centerline. At this level, the response is ruler flat from 100 Hz, the arbitrary crossover frequency setting down to 30 Hz. Subjectively, this was plenty loud. I measured the amplifier output voltage and read a minuscule 0.7 rms V on each channel. With a nominal 4.7-ohm impedance, this amounts to about 0.1 W per channel!

In Figure 3-4b, I increased the amplifier output until the sound pressure level reached a solid 100 dB. At this point you could begin to hear parts of the vehicle begin to resonate, which showed up on the spectrum analyzer as 3-dB ripple in the system response. The measured output voltage was still only 1.5 rms V or about 0.5 W per channel.

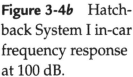

Figure 3-4b Hatchback System I in-car frequency response at 100 dB.

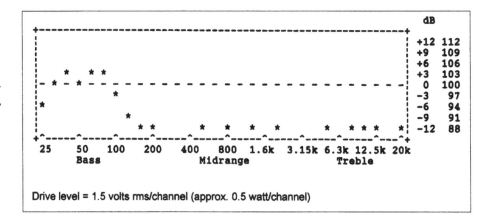

At higher levels, there was too much vehicle resonance to get meaningful spectrum data, so I just recorded SPLs during brief sound bursts. At 9.8 V (about 20 W) on each channel, the SPL was 115 dB. At this point, port noise began to be audible.

Finally, we took some oscilloscope waveform shots to see how clean the system looked at a few selected frequencies. We kept the drive level at 95 dB (still fairly loud) to minimize vehicle resonance. We performed this test with the filter out of the circuit. The results are shown in Photo 3-22a through d. In Photo 3-22a, the 30-Hz waveform looks somewhat triangular, but sounded fairly smooth. Some extraneous noises could be heard from the vehicle that disappeared by 40 Hz as seen in Photo 3-22b. At 50 Hz, a rather large resonance was excited and the waveform got a little ugly as seen in Photo 3-22c. At 100 Hz, there were no audible buzzes and the waveform became sinusoidal again.

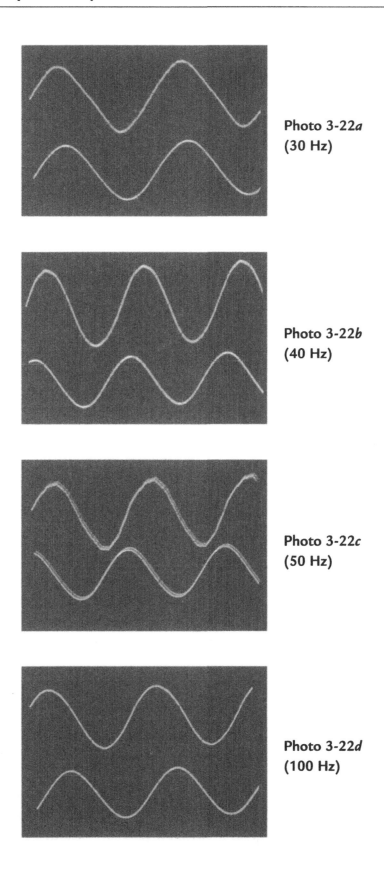

Photo 3-22*a*
(30 Hz)

Photo 3-22*b*
(40 Hz)

Photo 3-22*c*
(50 Hz)

Photo 3-22*d*
(100 Hz)

As a point of comparison, the predicted response in free air is shown as Figure 3-5*a*. Predicted filter settings in this environment would then be R1 = 5.48K and R2 = 87.7K, which provides a 6-dB boost at 33 Hz. These settings should provide flat response in larger volume spaces, like a large van or home application. Actual free-air response is shown as Figure 3-5*b*. The slight dip at 40 Hz is somewhat of an anomaly and could be due to the measurement environment or the fact that the box is 30 in above the floor.

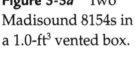

Figure 3-5*a* Two Madisound 8154s in a 1.0-ft³ vented box.

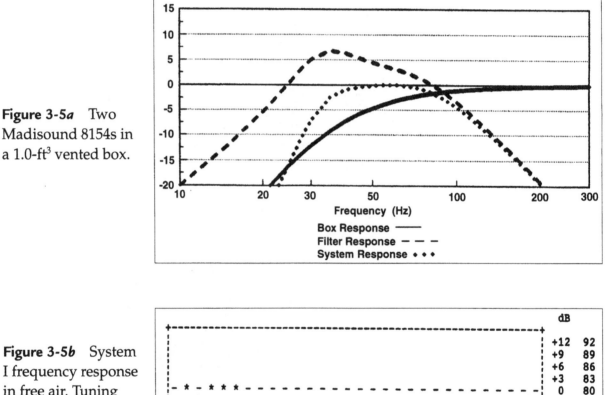

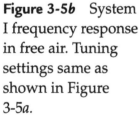

Figure 3-5*b* System I frequency response in free air. Tuning settings same as shown in Figure 3-5*a*.

Subjectively, this little 1-ft³ subwoofer is clean and powerful. It sounds great in a car or home application. It is simple to build and gives predictable results. Without a doubt, it's my favorite.

System II

Figure 3-6 shows the arrangement and details of a full-range hatch-back system that employs the best of both worlds—the compact, killer subwoofer above plus a pair of high-quality coaxes of your choice. In this case I chose the Madisound 5402 5-in coaxes with Audax hard dome tweeters for its solid midbass attack and extended highs. This system is only slightly more difficult to build than System I. The interior compartments, which isolate the coax, are made from two relatively small pieces. To better accommodate them, I replaced the single 3-in port with two 2-in ports. Equivalent area is only slightly less, so performance is not compromised. Again the box frequency is tuned to 33 Hz and the filter settings, if one is used, are the same as for System I above.

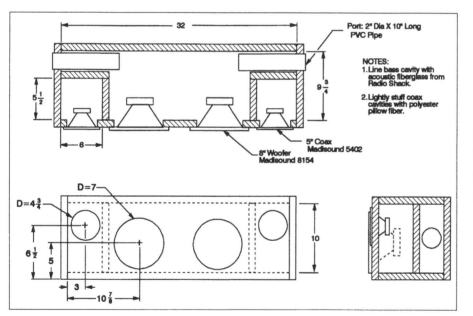

Figure 3-6 Hatch-back System II 8-in full-range rear speaker system.

The tools required to build System II are the same as those for System I.

Materials required are the same as for System I plus the following:

- 8 (16 total) No. 6 panhead sheet-metal screws, ¾ in long
- 2 ft of 2-in Schedule 40 PVC pipe
- 2 Madisound 5402 coaxes with grilles
- 2 (4 total) terminal cups

- 4 (8 total) ft of No. 16 or 18 speaker wire
- I package of polyester pillow fiber fill material

The parts used in this system are shown in Photo 3-23.

Photo 3-23

As always, we begin by cutting out all of the pieces to size. Photo 3-24 shows the two preassembled partitions clamped in place and ready for attachment to the bottom.

Photo 3-24

After the top is attached, the back is test-fit in as shown in Photo 3-25. In Photo 3-26, the back is attached and glue is applied to the mating surfaces for the top. Again, note the liberal quantity. This will ensure that there is plenty to be soaked into the end pores and fill minor joint imperfections.

Photo 3-27 shows the partial assembly from the back. From this position the shape of the woofer cavity can be seen. All that are missing are the back and end pieces, which are now in place in Photo 3-28. Again, note the overlapped edges which will be trimmed off with the router after the glue is completely cured.

Photo 3-25

Photo 3-26

Photo 3-27

Photo 3-28

In Photo 3-29, the woofer compartment has been lined with fiberglass and the cabinet has been placed facedown in preparation for installation of the four terminal cups and wiring. Photo 3-30 shows the terminal cups screwed into place with the No. 6 by ¾-in panhead sheet-metal screws.

Photo 3-29

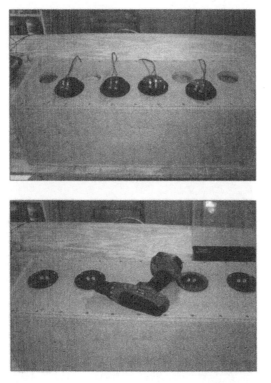

Photo 3-30

In Photo 3-31, all of the wiring is in place and the coax compartments have been lightly stuffed (but completely full) with polyester fiberfill. Please note that you will have to drill two small holes in the rear of the coax partitions to feed the wiring through prior to stuffing. These should be sealed with silicone prior to stuffing.

Photo 3-31

Photo 3-32 shows the ports now in place. At this point, you may choose to wait until the silicone has cured before proceeding. This really depends on how snug the holes for the ports are. If the ports will be disturbed during the rest of the installation process, it's best to let the caulk cure for several hours.

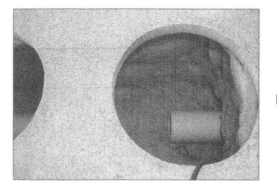

Photo 3-32

Photo 3-33 shows the completed system without the grilles installed. In Photo 3-34, the cabinet is placed in the well of the Firehawk used to test System I, so bass results should be virtually identical.

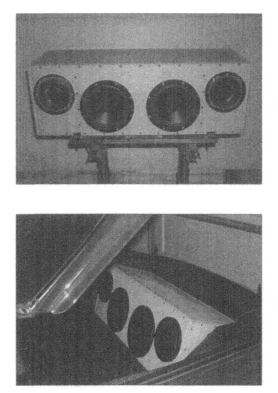

Photo 3-33

Photo 3-34

Moving on to the test data, Figure 3-7*a* shows the in-car frequency response taken at a sound pressure level of 90 dB with the mike positioned at the driver's seat, facing to the rear at about ear level. The overall response is 30 to 15 kHz ± 3 dB. The system developed 115 dB on pink noise with a total input of about 65 W with everything set for flat response.

Figure 3-7*a* Hatchback System II in-car frequency response.

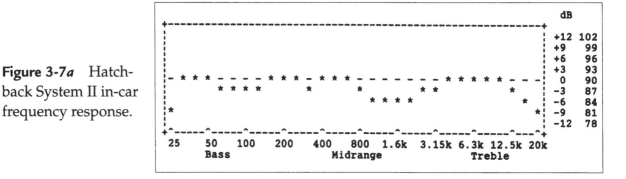

Subjectively, the sound has tremendous presence, in spite of the depressed midrange indicated in the response curve. Pop and rock music come alive reminiscent of listening to a stage monitor. Based on personal experience, these coaxes do a great job at the low levels typically associated with rear speakers—providing top end "air" and a sense of increased cabin space. As we would expect, the bass is smooth and extended, identical to System I.

For comparison, I took some data in free air to eliminate the vehicle effects. The result is shown in Figure 3-7*b*. The mike was positioned on the center axis of one of the coaxes at a distance of 18 in. This reduced octave-to-octave variability somewhat and captured more of the treble energy. After that, I hooked this system up to my

Figure 3-7*b* System II response in free air. Microphone positioned on coax center at distance of 18 in.

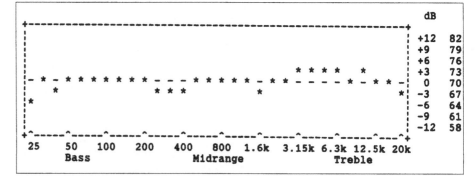

home system and listened to it with a variety of music. Overall, the sound was fairly pleasant. The coax is a bit too forward for my tastes, however.

One final point. It should be noted that we could have done a lot more to make this cabinet and System I more aesthetically suitable for this particular vehicle. (For openers, I left off the carpeting specifically to get more detail in the photographs.) But if I were designing a cabinet specifically to fit the Firehawk, it would probably not be rectangular. There are lots of possibilities other than a square box that we could envision to make Systems I and II blend in better with a specific environment. In the final analysis we must balance ease of construction with personal ability and availability of power tools. These two are only intended to provide insight as to what is required to properly construct a high-quality speaker system.

More Power

Closed-Box System

This chapter would be incomplete without constructing at least one higher-powered subwoofer. For our example, we chose a system with a pair of 10-in woofers—modest compared to say . . . a pair of 15s, but still quite powerful. Again, major consideration was given to size, and this one is a fairly compact two-cubic-footer. The dimensions worked out so that you can make this cabinet from 11½-in shelving without having to make any rip cuts. (All the pieces are 11½ in wide.) We're going to explore several different variations using this same cabinet and drivers, which will give you some choices to experiment with.

Our first example is a closed box—by far the easiest to build and the most forgiving. The cabinet is shown in Figure 3-8.

For this project we will need the following materials:

- ¾-in-thick particle board—two 8-ft lengths of 11½-in-wide shelving
- Box of coarse-thread drywall screws, 1⅝ in long

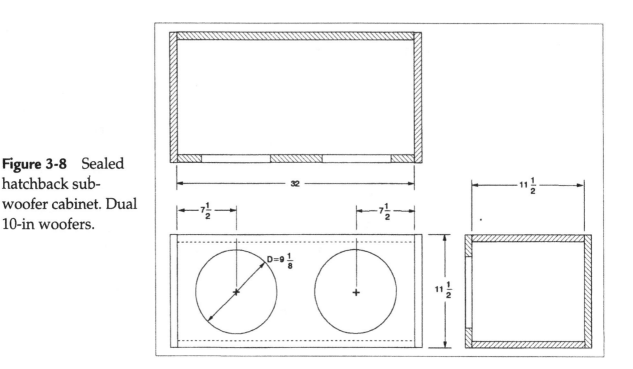

Figure 3-8 Sealed hatchback sub-woofer cabinet. Dual 10-in woofers.

- Yellow carpenter's glue
- Small box of rope caulk
- 8 No. 6 panhead sheet-metal screws, ¾ in long
- 3 20-oz packages of polyester fiberfill
- 2 Madisound 10207 10-in woofers with grilles
- 2 terminal cups
- 3 ft of No. 16 or 18 speaker wire
- Solder

The components are shown in Photo 3-35. These are beefy drivers with 40-oz magnets and dual 8-ohm voice coils, each with a power rating of 100 W.

Photo 3-35

We've shown the construction techniques numerous times in the past and won't be repeating them here. The assembled cabinet is shown in Photo 3-36 and again in Photo 3-37 after trimming the edges flush with the router and rounding them over with a ⅜-in-radius roundover bit. As you can see, the improvement is subtle, so if you don't own a router you could opt to leave that step off. Alternatively, with some patience and effort you could even do something similar manually with a good wood rasp.

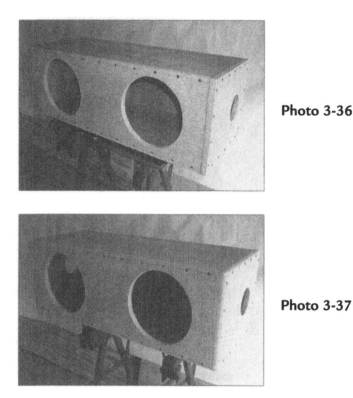

Photo 3-36

Photo 3-37

In Photo 3-38, the terminal cups and wiring have been installed on the back and the 60 oz of fiberfill are in place. In Photo 3-39 the drivers and grilles are now screwed into place.

Note: In this first setup, we will be connecting only one of the two 8-ohm voice coils on each woofer in order to increase the Q to make this driver more suitable for a closed-box application. In so doing, the actual measured Q_{TS} exactly doubled from the advertised value of 0.22 to 0.44. Combining this with the low measured free-air resonant frequency of 17 Hz (advertised as 19) makes this woofer now nearly ideal for a closed-box application. To get maximum amplifier power transfer, you'll probably want

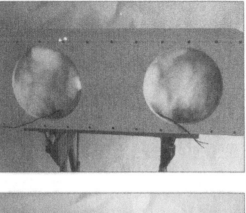

Photo 3-38

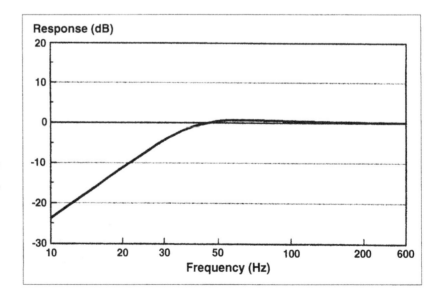

Photo 3-39

to connect the two woofers in parallel to get a nominal 4-ohm rating and then set your amplifier to mono-bridged mode.

Figure 3-9a shows the predicted closed-box response in free air. It is down 3 dB at a very respectable 34 Hz. The only negative is a predicted hump of 1.6 dB in the response curve.

Figure 3-9a Predicted free-air frequency response. Two Madisound 10207 10-in woofers in a 2-ft³ closed box. One voice coil per driver connected.

As a first test, I powered the system to an SPL of 90 dB at 1 meter with a full-range pink noise signal and measured the response in free air. The result is shown in Figure 3-9*b*. It would appear to be very close to the model—which is one of the main attributes of closed-box systems. They are very predictable.

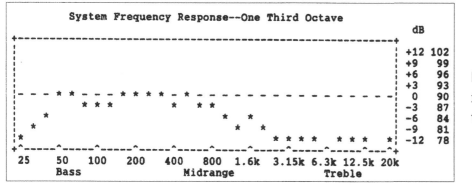

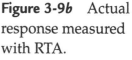

Figure 3-9*b* Actual response measured with RTA.

In-car test results are shown in Figure 3-10*a* and *b*, both at an SPL of 90 dB. The low-end reinforcement provided by the small volume of the vehicle cabin is apparent in both curves.

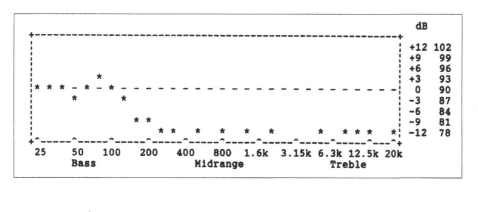

Figure 3-10*a* In-car response for dual 10-in closed-box system driven by Jensen 50-W-per-channel amplifier. Response with amplifier's built-in low-pass filter.

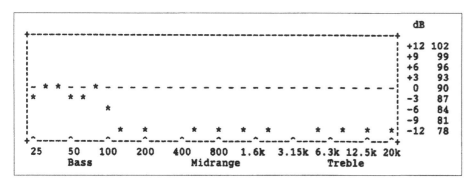

Figure 3-10*b* In-car response for dual 10-in closed-box system driven by Jensen 50-W-per-channel amplifier. Response with subwoofer filter settings: no low-frequency boost and 30-Hz cutoff.

In Figure 3-10a, the system is powered by a 50-W-per-channel amplifier with a built-in subwoofer filter that appears to have a low-pass corner frequency of about 125 to 150 Hz. Low-bass extension is flat all the way off the left of the chart. It's possible that this system would sound satisfactory with no other signal-conditioning components.

For comparison, I connected my subwoofer filter set for flat response with a 30-Hz cutoff to minimize power loss in the subaudible region. As seen from the graph, the system is still down only 3 dB at 25 Hz.

At full amplifier power of 100 W rms, the system generated an in-car SPL of 115 dB. I then connected a 100-W-per-channel amplifier which generated 120 dB. It took the power effortlessly.

Subjectively, this system is powerful, clean, and tight—again, characteristic of closed-box systems. I found it quite musical and having plenty of headroom. All in all, if you can afford to give up 2+ ft^3 of space, this is an easy system to build and get predictable results. If your available space won't accommodate a box this large, you could make a box half this size and install a single 10207, which would probably develop about 112 dB.

Vented System with Dual 3-In Ports

With both voice coils connected, the driver Q_{TS} drops to its advertised value of 0.22, making the Madisound 10207s prime candidates for a vented system. In general, vented systems have lower bass extension and can handle higher power at low frequencies since the box and port are doing most of the work. In fact, at box resonance frequency, which is typically around 30 Hz, the woofer cones almost stop moving, which means you can pour on the power until port noise becomes too intrusive. The diameter of the port determines how loudly the system can play without generating port noise—the larger the better.

In our first example, the box is tuned to 30 Hz to accommodate two 3-in-diameter ports placed on opposite sides of the cabinet. This makes it perform better under the seat of some full-sized vans and distributes the low-frequency energy more evenly throughout the passenger space. Figure 3-11 shows the general arrangement. The actual placement of the ports is noncritical provided they are located a minimum of approximately 1½ in from any interior box surface and that they each have 3 in of exterior clearance from any vehicle structure.

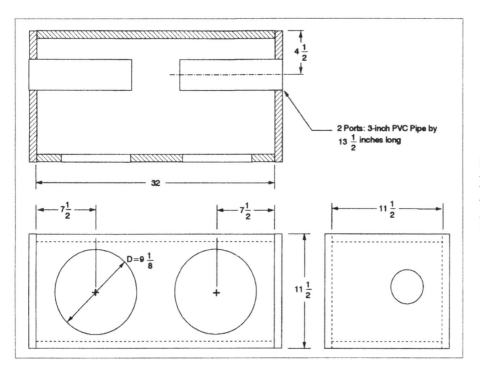

Figure 3-11 Vented hatchback subwoofer cabinet. Dual 10-in woofers.

Photo 3-40 shows the ports installed and the cabinet lined with Radio Shack fiberglass. In Photo 3-41, one of the woofers is shown lying face down with the dual voice coils wired in parallel giving them a nominal 4-ohm impedance rating. Due to the increased size of this driver, I elected to install three rows of rope caulk on the back of the grille ring instead of the usual two.

Photo 3-40

Photo 3-41

The predicted free-air response of this system is shown in Figure 3-12*a*. From the size of the hump in the response, it would appear to be mistuned—and it actually is. The measured free-air response confirms this in Figure 3-12*b*. The box frequency was measured to be nearly exactly 30 Hz and the subwoofer filter was set for the theoretical values—$Q = 3.0$ at 24 Hz (R1 = 5K, R2 = 181K). The two curves appear to be somewhat similar.

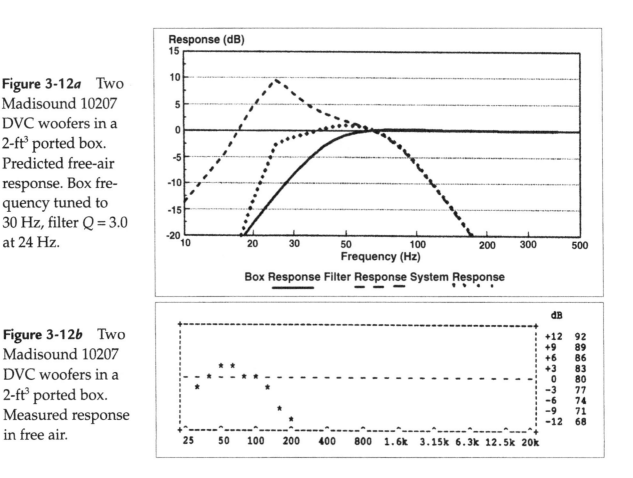

Figure 3-12*a* Two Madisound 10207 DVC woofers in a 2-ft³ ported box. Predicted free-air response. Box frequency tuned to 30 Hz, filter $Q = 3.0$ at 24 Hz.

Figure 3-12*b* Two Madisound 10207 DVC woofers in a 2-ft³ ported box. Measured response in free air.

The in-car measured response is significantly different from free air, which we saw earlier. The typical vehicle space causes a boost in the lowest frequencies. For this environment, the filter settings were adjusted to R1 = 17K and R2 = 34K, corresponding to a flat filter response (Q = 0.7) and our standard 30-Hz cutoff for rumble suppression. The result is Figure 3-12c. In this particular vehicle, the response is flat from our arbitrary 100-Hz crossover setting out past 25 Hz, the analyzer measurement limit. Admittedly, part of this is due to luck, at least regarding the exact combination of vehicle size and shape.

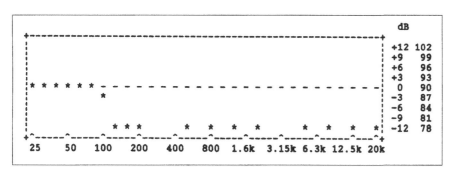

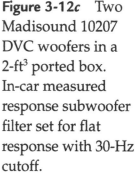

Figure 3-12c Two Madisound 10207 DVC woofers in a 2-ft³ ported box. In-car measured response subwoofer filter set for flat response with 30-Hz cutoff.

Power handling of this system exceeded the 200 W I had available to drive it. The voice coils are rated at 100 W each, making this system thermally capable of handling 400 W. As with the closed-box system, at full power it generated 120 dB but was amplifier limited. With the addition of the rumble filter, power handling should be around 400 W. Even so, theoretically, this is an increase of only 3 dB.

Subjectively the sound is similar to the closed-box version, but I can hear deeper harmonics present. I don't think the sound is quite as tight as the sealed version.

Vented System with One 4-In Port

If you want to experiment with even lower frequencies, you can try our last version, which is shown in Figure 3-13. Instead of the dual 3-in ports, we now have a single pipe 20 in long and 4 in in diameter. The advantage here is that we can now tune this box to the theoretical free-air target frequency of 23 Hz or anything in between.

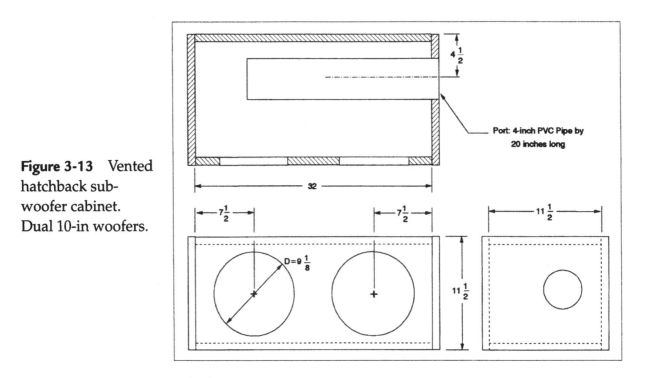

Figure 3-13 Vented hatchback sub-woofer cabinet. Dual 10-in woofers.

Photo 3-42 shows what the big port looks like. The one in this photo was set to the theoretical length and is actually longer than the final value, which was determined experimentally to be 20 in. From a practical standpoint, a port this long should be supported at both ends. This is easily done by resting the pipe on a block glued to the rear wall and then attaching the pipe to the block with a generous blob of silicone.

Photo 3-42

Figure 3-14*a* shows the predicted free-air response. The theoretical subwoofer filter setting is a *Q* of 2.7 at 25 Hz, which corresponds to R1 = 5.4K and R2 = 156K.

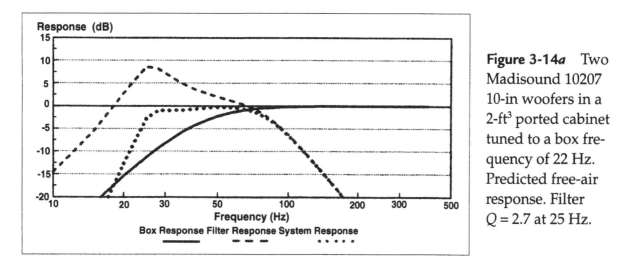

Figure 3-14*a* Two Madisound 10207 10-in woofers in a 2-ft³ ported cabinet tuned to a box frequency of 22 Hz. Predicted free-air response. Filter *Q* = 2.7 at 25 Hz.

The measured free-air response is shown in Figure 3-14*b*. However, to achieve this, the filter settings had to be significantly altered from theoretical, indicating some problem with the data. Actual filter settings were *Q* = 3.97 at 31.4 Hz corresponding to R1 = 2.9K and R2 = 183K, which is probably indicative of the box frequency being set too low. With a deviation this large, I hesitated to include this version in the book. However, it does illustrate a point. Many authors have written about the difficulty in achieving proper alignment of ported systems. This one is illustrative of the problems which can sometimes be encountered.

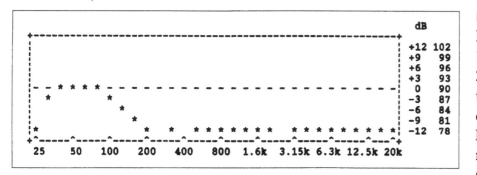

Figure 3-14*b* Two Madisound 10207 10-in woofers in a 2-ft³ ported cabinet tuned to a box frequency of 22 Hz. Measured free-air response. Filter *Q* = 3.97 at 31.4 Hz.

In spite of all this, the in-car response was still respectable and filter settings were much more reasonable with $Q = 1.43$ at 34.7 Hz, corresponding to R1 = 7.3K and R2 = 59.6K. The results are shown in Figure 3-14c. This system may sound a little tighter than the dual-ported system above. Theoretically, the power handling should be greater due to the larger port diameter. On the negative side, the single port may be difficult to position in the vehicle to get uniform bass energy distribution.

Figure 3-14c Two Madisound 10207 10-in woofers in a 2-ft³ ported cabinet tuned to a box frequency of 22 Hz. Measured in-car response. Filter $Q = 1.43$ at 34.7 Hz.

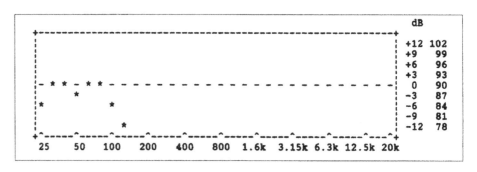

Which one of the three 10-in systems should you build? Clearly the sealed version is the easiest and most predictable. The dual-port vented version should work in many applications but may not sound tight enough for some tastes. Lastly, the single-ported version may be better suited for home applications or larger spaces like vans and motor homes.

In Summary

One final comparison before closing this chapter. Photo 3-43 shows the 1 ft³ dual 8-in system stacked on top of the 2-ft³ dual 10-in system. For listening to most music, the increased size may not be worth the back strain or loss of cabin space. However, for die-hard bassheads, the choice is clear. Go with the 10s.

Photo 3-43

Chapter 4

Rear Systems
for Pickup Trucks

Standard Cab

After the sedan, the next most difficult design problem in auto sound is how to get good bass in a standard pickup truck. The main speakers usually aren't much of a problem, since pickup trucks can often have more space available in the doors and dash than passenger cars do. This makes installation of upgraded components in these areas easier and more straightforward. But the space available for a bass cabinet is minimal and somewhat difficult to utilize effectively.

For many years, the standard pickup truck box was a wedge-shaped affair, placed behind the seat, with the woofer essentially in contact with the seatback. Photo 4-1 is typical of this type. What

Photo 4-1

happens with this arrangement is that most of the sound is absorbed into the seat . . . and you, the passenger! In a high-powered system, this literally feels like a series of kidney punches. For some program material, the pitch of the bass notes is difficult to discern, and they simply become a series of muffled thuds. Thankfully, a few manufacturers have seen the light in recent years and are making some changes to improve on this situation. However, Photo 4-1 is still the rule and not the exception when looking for a commercial solution to the pickup truck bass problem.

Good News

There *is* a better way to deal with this problem. The solution is shown in Figure 4-1. In this arrangement, the bass drivers are positioned to fire downward so that the sound is projected into the space between the floor and seat bottom. This provides effective coupling to the air mass inside the truck instead of your kidneys and makes for a huge improvement in the pickup truck audio experience. In fact, the quality of the bass in this arrangement will be as good as in any other type of vehicle. The trick is to fit this all into a very compact space without affecting front-to-rear seat travel, which many of us would find intolerable.

Referring again to Figure 4-1, our standard cab pickup truck subwoofer cabinet has a trapezoidal cross section, which transitions into a rectangular cross section. This shape mirrors the offset between the seatback and seat bottom (if such an offset exists) and varies in size from vehicle to vehicle. In most installations, the space is very restricted. So, without the rectangular section, it would not be possible to accommodate 8-in woofers, which we consider to be the minimum. If the offset is not needed, the cabinet then becomes a straight trapezoid.

The formula for calculating the internal volume is also given in the diagram. It is simply the sum of the areas of the trapezoid and rectangle multiplied times the cabinet width.

We begin the design of our standard cab subwoofer cabinet by taking measurements and then transferring them to a full-sized tem-

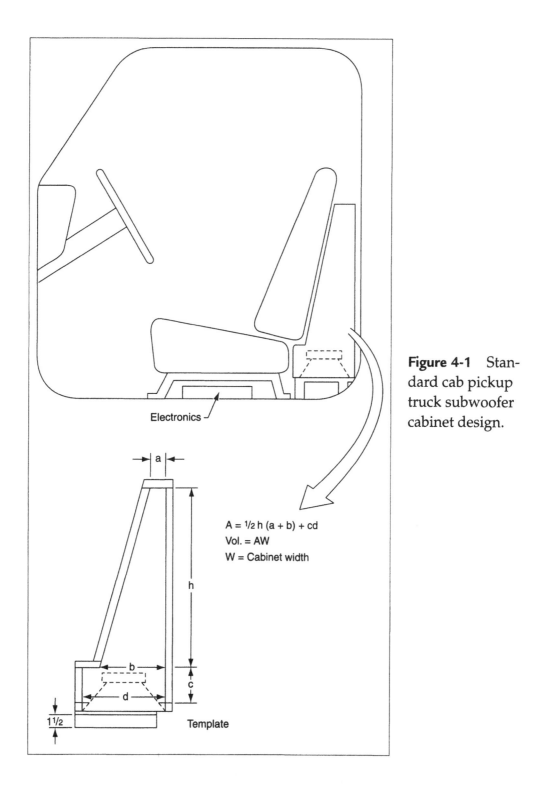

Electronics

$A = \frac{1}{2} h\,(a + b) + cd$

Vol. = AW

W = Cabinet width

a

h

b

c

d

$1\frac{1}{2}$

Template

Figure 4-1 Standard cab pickup truck subwoofer cabinet design.

plate, which will duplicate the cross-sectional area of the subwoofer cabinet. In Photos 4-2 and 4-3, the seat has been removed from the test vehicle (a 1988 Toyota extended cab) and placed on the worktable for ease of illustration.

Photo 4-2

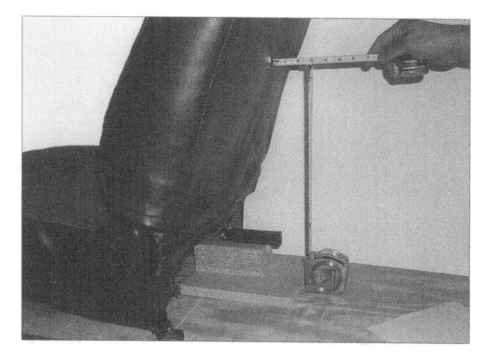

Photo 4-3

We start the process by positioning the seatback at the greatest angle we're likely to need and design our cabinet slope to accommodate that. Assuming that the truck cab has a vertical rear wall, we positioned some convenient blocks of scrap, and a tape measure to outline the shape. The dimensions taken are transferred to poster board which becomes a full-scale drawing as seen in Photo 4-4. These are used to develop the construction drawing shown in Figure 4-2. In our test vehicle, the total clearance between the seat bottom and the floor is only 3 in. In this case, dimension c in Figure 4-1 will be zero. In other words, the rectangular box section will consist only of an overlap between the speaker baffle board and sidewalls. This makes it a little more difficult to get adequate clearance between the sloped front cabinet wall and the woofer magnet and frame. We'll show more detail on this later on.

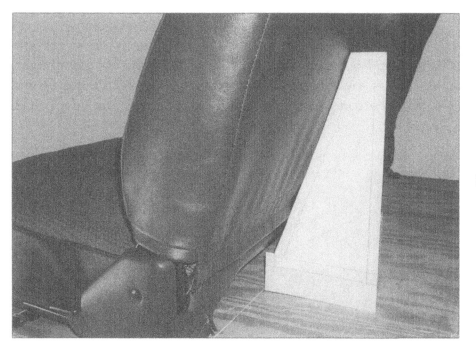

Photo 4-4

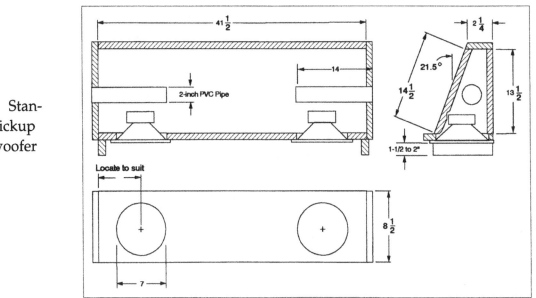

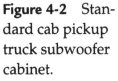

Figure 4-2 Standard cab pickup truck subwoofer cabinet.

After cutting all the pieces to size, we begin the assembly process by attaching the top to the front with a few screws as shown in Photo 4-5. In Photo 4-6, the ends are attached. With the aid of a pair of right-angle clamps, the bottom is fit into place in Photo 4-7. Photos 4-8 and 4-9 show the completed assembly after trimming the edges flush

Photo 4-5

Photo 4-6

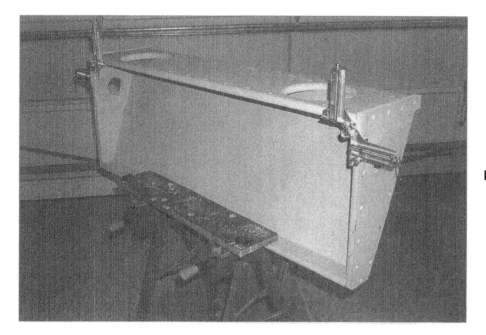

Photo 4-7

Photo 4-8

Photo 4-9

with the router. In Photo 4-10, the close-up of one of the woofer mounting holes shows how the walls encroach into the woofer mounting space. This area must be relieved by filing or grinding with something like a Dremmel tool. The result can be seen in Photo 4-11. Also in this photo, the terminal cups, wiring, and 14-in-long, 2-in-diameter ports have been installed.

Photo 4-10

Photo 4-11

System Tests

After some 2-in spacers are installed to act as feet, the completed system is shown placed behind the truck seat to verify fitup in Photo 4-12. We next measured the box resonance frequency and found it to be 28.5 Hz, just about on target. After that, we ran some response curves with the real-time analyzer.

Photo 4-12

Figure 4-3*a* shows the response in free air as the result of feeding pink noise into the test amplifier and using its on-board low-pass filter. It's reasonably flat, and although it has dropped by 9 dB at 40 Hz, we have seen numerous times that the vehicle will add quite a bit of lift at the low end. But again, this test amplifier crossover point looks too high—around 200 Hz. In Figure 4-3*b*, the onboard filter is replaced by our subwoofer filter tuned to a Q of 2.0 at 33 Hz, corresponding to $R_1 = 5.48K$ and $R_2 = 87.7K$. The measured free-air response is outstanding—flat all the way out to 31.5 Hz with the rapid drop-off needed for good rumble suppression. This curve is actually more favorable than the theoretical free-air response shown in Figure 4-3*c*. It's flatter and has better low-end extension by about 3 Hz.

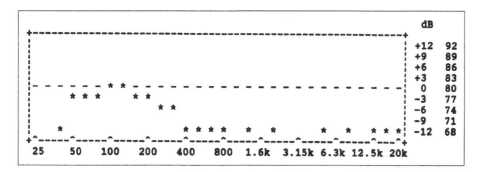

Figure 4-3*a* Standard cab pickup truck subwoofer frequency response. Two Madisound 8154 8-in woofers in a 1-ft³ vented box. Free-air response with test amplifier on-board crossover.

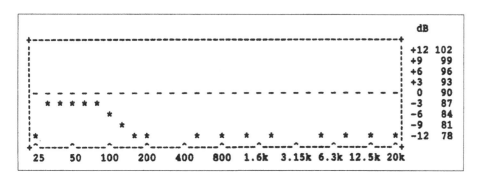

Figure 4-3*b* Standard cab pickup truck subwoofer frequency response. Two Madisound 8154 8-in woofers in a 1-ft³ vented box. Free-air response with subwoofer filter set for $Q = 2$ at 33 Hz.

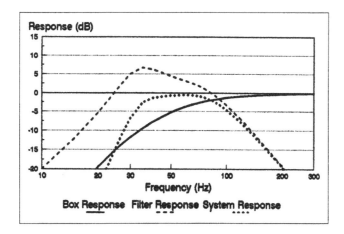

Figure 4-3*c* Standard cab pickup truck subwoofer frequency response. Two Madisound 8154 8-in woofers in a 1-ft³ vented box. Theoretical free-air system response.

In-car testing results were also favorable. To characterize the vehicle space, we ran a full-range signal to the system and got the results shown in Figure 4-4*a*. Again, we see the low-bass lift, extending the frequency response off the chart. On the negative side, this particular space has a 6-dB notch at 80, followed by a 6-dB peak at 100 Hz. Since we needed no low-end boost, we set the subwoofer filter for a flat response and a 30-Hz cutoff—our standard settings for good rumble suppression—and got the response shown in Figure 4-4*b*. While this

Figure 4-4*a* Measured in-vehicle response. Standard cab pickup truck subwoofer system. Two Madisound 8154 8-in woofers in a 1-ft³ vented box. Full-range response.

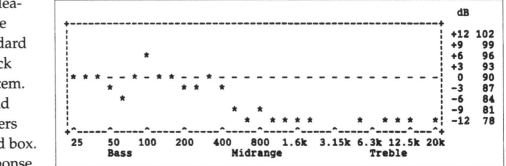

Figure 4-4*b* Measured in-vehicle response. Standard cab pickup truck subwoofer system. Two Madisound 8154 8-in woofers in a 1-ft³ vented box. Response with subwoofer filter.

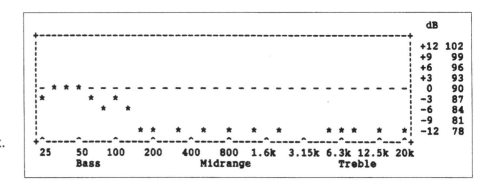

is an acceptable curve, the 80-Hz dip, which remains, is undesirable since it has a negative effect on midbass attack. In listening to music through this system, I was able to minimize this problem by adjusting the subwoofer crossover frequency.

Finally, since this particular truck is equipped with a 250-W amplifier, I tested the sub's power-handling capability and sound pressure level. For very brief periods, the system developed 122 dB on pink noise at about a 200-W input. Based on past experience, this system can develop peaks of about 125 dB in a pickup truck with music as a program source.

Other Possibilities

What we have shown thus far is one example of an improved subwoofer system for the standard cab pickup truck. Personal experience has shown that this particular design provides a reliable solution to the problem. However, there are some other possibilities. For instance, we would choose a somewhat different approach for an extended cab truck, and if more power handling is needed, we need more cone area than can be provided by just two 8-in drivers.

Figure 4-5 shows six rear speaker layouts for both standard and extended cab trucks. Although the front and rear main speakers are shown as a visual aid, we are going to assume that whatever drivers are there are mounted in the stock locations and have been upgraded as necessary. So for now, the discussion is focused on the design of the subwoofer system.

Standard Cabs: Systems I to III

Pickup System I is a compact design employing two 6½-in drivers. The only purpose for using the 6½s instead of our standard 8s is to make the system fit into the narrowest space possible—like that found in the Ford Ranger. The problem is that there are few, if any, 6½-in drivers suitable for use as a subwoofer, so this is a significant

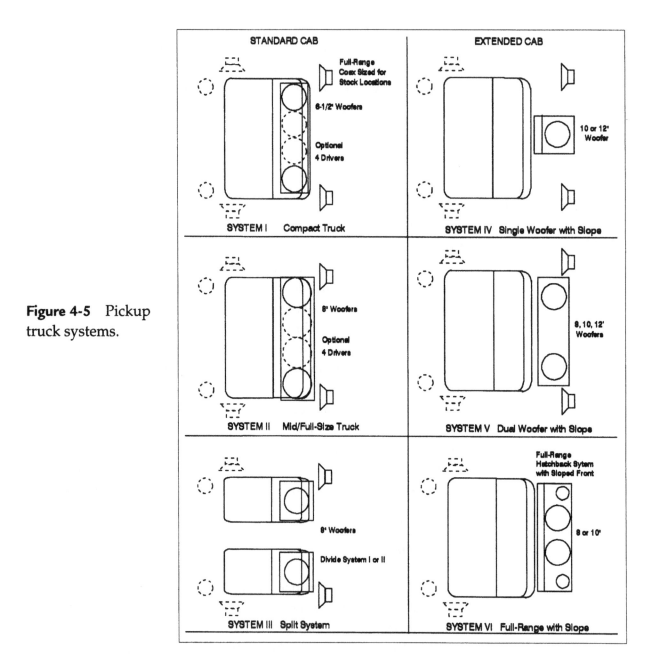

Figure 4-5 Pickup truck systems.

compromise. Our choice here is to use two Madisound 6102-4s in a 0.6-ft³ enclosure, which can be of either sealed or vented design. While this system performs OK, it is the least desirable of those in this section as box size has been reduced to much less than optimum and low-end boost is higher than it is for the other systems. Predicted subwoofer frequency responses are shown in Figure 4-7a and b.

The limited power-handling ability of the sealed version makes the vented version a more attractive alternative. Also, both versions greatly benefit by scaling up to four drivers, which should be possible even in the Ranger. Four 6s have approximately the same cone area as a single 12, which should be plenty to drive the space in a compact pickup truck.

System II is the system we built for demonstration purposes—a somewhat scaled up version of System I employing two 8-in Madisound 8154s in a 1.0-ft³ vented enclosure. It will fit into virtually any compact truck (except the Ford Ranger—according to one of my clients). As stated earlier, this is a terrific-sounding compact subwoofer system, and its performance in any pickup is nothing short of spectacular.

If you need very high sound pressure levels, you can expand the cabinet to accommodate four 8154s. This provides an effective cone area greater than two 10s but somewhat less than two 12s. At this increased size, however, you may have to give up some seat travel to get it to fit. Considering the small volume of a conventional cab pickup, this one ought to have enough output to peal the paint.

System III is a split version of System II, which is configured for trucks with bucket seats. All of the designs in Systems I and II are symmetric and can therefore simply be divided in half.

Figure 4-6 summarizes design data for Systems I, II, and III, with corresponding theoretical free-air responses shown in Figure 4-7. We have seen repeatedly that the typical vehicle space will impart a rising low-frequency response, which means less boost will be required than indicated. It follows that these designs have the potential to work fairly well without my subwoofer filter, provided that some other type of electronic crossover is used to filter out high-frequency content. The closed-box system based on the Madisound 81524 DVC (with both voice coils driven) looks the most promising to generate extended low-frequency response without the aid of electronic boost. This driver also has greater linear excursion capability than the others in this group, making it better suited to closed-box applications.

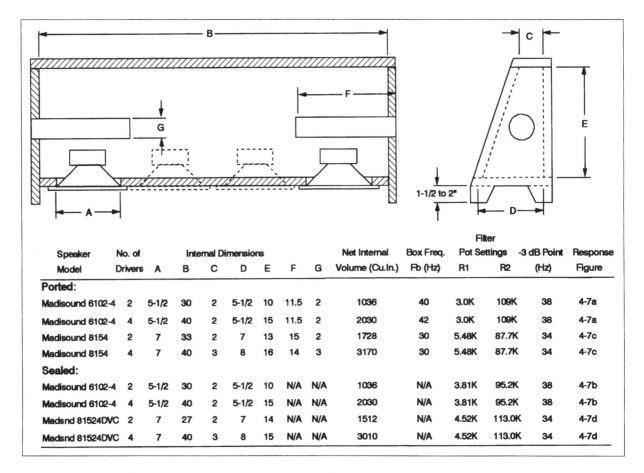

Speaker Model	No. of Drivers	Internal Dimensions							Net Internal Volume (Cu.In.)	Box Freq. Fb (Hz)	Filter Pot Settings		-3 dB Point (Hz)	Response Figure
		A	B	C	D	E	F	G			R1	R2		
Ported:														
Madisound 6102-4	2	5-1/2	30	2	5-1/2	10	11.5	2	1036	40	3.0K	109K	38	4-7a
Madisound 6102-4	4	5-1/2	40	2	5-1/2	15	11.5	2	2030	42	3.0K	109K	38	4-7a
Madisound 8154	2	7	33	2	7	13	15	2	1728	30	5.48K	87.7K	34	4-7c
Madisound 8154	4	7	40	3	8	16	14	3	3170	30	5.48K	87.7K	34	4-7c
Sealed:														
Madisound 6102-4	2	5-1/2	30	2	5-1/2	10	N/A	N/A	1036	N/A	3.81K	95.2K	38	4-7b
Madisound 6102-4	4	5-1/2	40	2	5-1/2	15	N/A	N/A	2030	N/A	3.81K	95.2K	38	4-7b
Madsnd 81524DVC	2	7	27	2	7	14	N/A	N/A	1512	N/A	4.52K	113.0K	34	4-7d
Madsnd 81524DVC	4	7	40	3	8	15	N/A	N/A	3010	N/A	4.52K	113.0K	34	4-7d

Figure 4-6   Standard cab pickup truck subwoofer cabinets.

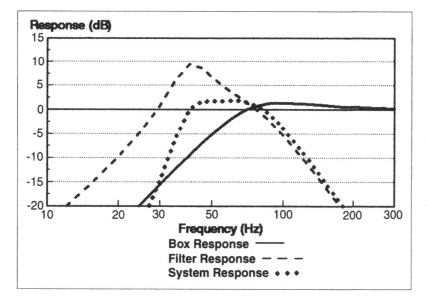

Figure 4-7a Two Madisound 6102-4s in a 0.6-ft³ vented box. $Q_f = 3.0$ at 40 Hz; R1 = 3.0 K; R2 = 109 K; $F_b =$ 42 Hz; F3 = 38 Hz.

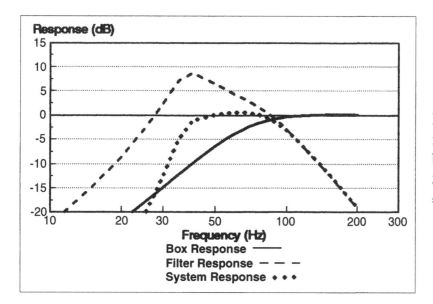

Figure 4-7b Two Madisound 6102-4s in a 0.6-ft³ closed box. $Q_f = 2.5$ at 35 Hz; R1 = 3.8 K; R2 = 95.2 K; F3 = 38 Hz.

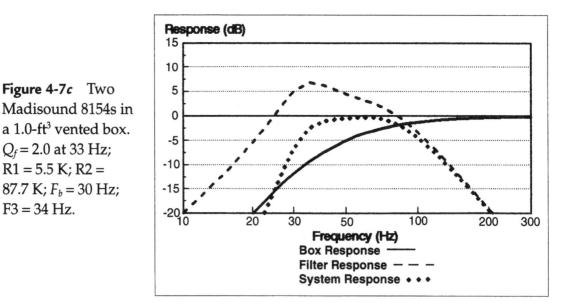

Figure 4-7c Two Madisound 8154s in a 1.0-ft³ vented box. $Q_f = 2.0$ at 33 Hz; R1 = 5.5 K; R2 = 87.7 K; F_b = 30 Hz; F3 = 34 Hz.

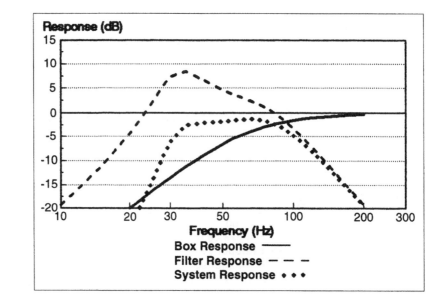

Figure 4-7d Two Madisound 81524DVCs in a 1.0-ft³ closed box. $Q_f = 2.5$ at 32 Hz; R1 = 4.5 K; R2 = 113 K; F3 = 34 Hz.

In general, four drivers are recommended for these closed-box systems. If space won't permit installing a system based on four 81524DVCs, you could probably get by with three drivers by scaling the enclosure down proportionally. I personally wouldn't consider putting in a closed-box system based on only two 6-in drivers.

Standard Cab System Design

Referring back to Figure 4-1, a, b, c, d, and h are the inside dimensions needed to calculate box volume. We saw previously that the first step in designing the cabinet is to carefully measure the shape of the space behind the seat and record the corresponding outside dimensions and the cab width W. Subtract 1½ in from each of these dimensions to account for the ¾-in thickness of the particle board and record these new figures as inside dimensions. Then, calculate the cross-sectional area A from the formula shown in the figure and multiply it times W to determine the available box volume. Be sure to allow for the 1½- to 2-in clearance required between the bottom of the speaker cabinet and the cab floor. In practice, floor clearance is usually provided by the driveline hump in the floor pan, so in some instances the cabinet could actually rest on the hump. If you are installing a ported system, allow 2 in at each end of the cabinet for the space needed for ports to fire into.

To be absolutely certain of your measurements, cut out a template from some scrap plywood and place it behind the seat. Once you confirm it fits and that the seat is as far back as you will ever want it, you are ready to proceed with system selection. If you find that you have to make some adjustments to the template, redo your calculations until you arrive at the final internal box volume.

Referring back to Figure 4-6, box cross-sectional shapes are assumed to be trapezoidal for comparison purposes. (In a pinch, they could be used as shown.) However, your task is to adjust either height h or box width W (or both) to make your stepped box volume equal one you select from Figure 4-6. After that, set port lengths and filter pot settings to correspond to those shown in the design you selected. This process may seem a little complicated at first but it's not really, and the results will be well worth the effort.

Do your best to attain the full, specified box volume. If you simply can't, make the box as large as possible. Probably, you won't miss it by more than 10 percent, which you should be able to overcome by filter tuning. An error greater than this is apt to be audible as a peak in the bass response.

Extended Cab Systems

System IV is an extended cab system with a single 10- or 12-in woofer facing either up or down in a trapezoidal, sealed enclosure. It is created by adding a sloped front to the single woofer designs given in Figure 6-4. To assist you in this, Figure 4-8 shows how to convert any of the rectangular box designs to a sloped front and provides the formula for calculating the volume of a trapezoidal enclosure. Another, and perhaps more attractive, alternative would be to design your own cabinet to exactly fit your vehicle using the data from Figure 6-4.

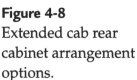

Figure 4-8
Extended cab rear cabinet arrangement options.

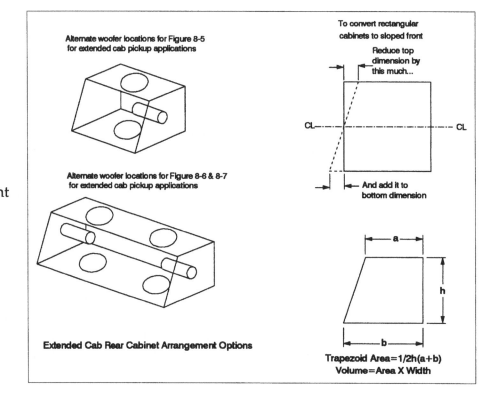

For best aesthetics and space utilization, the slope should be set somewhat parallel to the seatback. If you use your truck for road trips, you'll probably want to allow enough space for full seat travel and for your passenger to stretch out.

When space permits, the 12-in driver version is recommended. See Figure 6-9a, 6-9d, and 6-10c for predicted responses.

The System V subwoofer can be any of the hatchback systems from Chapters 3 and 6 which employ two 8-, 10-, or 12-in woofers in a variety of alignments. Again, for a more attractive and practical installation, the cabinets should have a trapezoidal cross section.

Owners of full-sized extended cab pickups will probably want to build something to fit the space under the rear seat. The 1-ft^3 system with the two Madisound 8154s we used in our example system in this chapter is your best bet here, with the woofers again firing downward.

I installed one of these in a 1999 Ford Lariat, and got great results. Photo 4-13 shows the completed cabinet lying on its back. Note the ports extending 2 in outside of the cabinet, which was required to get sufficient end clearance inside the box. Note also the steps and notches needed to utilize every cubic inch of space possible. In spite of all this, the cabinet netted out only 0.7 ft^3, but we were able to tune around it using modified filter settings. The predicted free-air response is shown in Figure 4-9.

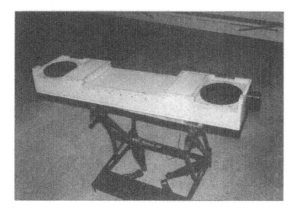

Photo 4-13

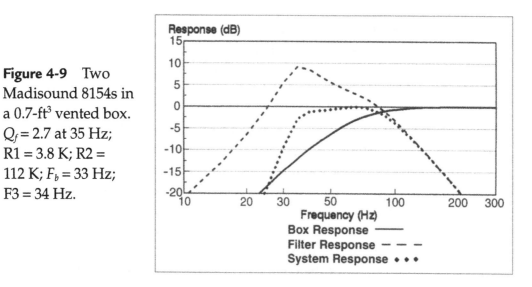

Figure 4-9 Two Madisound 8154s in a 0.7-ft³ vented box. Q_f = 2.7 at 35 Hz; R1 = 3.8 K; R2 = 112 K; F_b = 33 Hz; F3 = 34 Hz.

Photo 4-14 shows the cabinet now carpeted and ready for installation. In Photo 4-15, the cabinet is installed and the port is clear of any obstructions, providing excellent coupling to the air space. Photo 4-16 is a shot at floor level showing how the cabinet conforms to the shape of the floor plan. All these space constraints made this is very

Photo 4-14

Photo 4-15

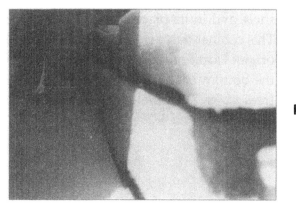

Photo 4-16

difficult cabinet to build, and it would have been nearly impossible to get it right without constructing a full-sized mockup out of styrofoam house sheathing—a new trick I learned on this project.

System VI assumes that no rear main speaker receptacles are available and uses one of the full-range hatchback systems from Chapters 3 and 6. Again, the sloped-front option shown in Figure 4-10 is recommended.

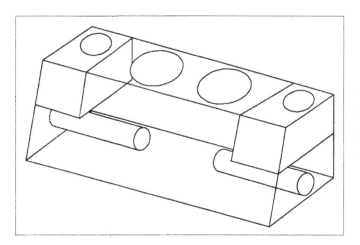

Figure 4-10
Extended cab pickup truck full-range rear cabinet.

Hatchback System II in Chapter 3, which we covered in detail, formed the basis for the system I have used for years in my personal vehicle, a 1988 Toyota extended cab. This particular truck has a relatively small cab extension compared to other makes. There is insufficient room for any kind of rear seat and, originally, it came with only a removable shelf in the back, which covered the jack. I removed the shelf, and in its place, I installed a modified version of Figure 4-10. This cabinet has a step in the bottom to accommodate the jack in its original location and a large bulge in the floor pan. Photo 4-17 shows the bottom of the cabinet where the woofers and port reside. This frees up most of the top surface to serve as a shelf again as seen in Photos 4-18 and 4-19.

Photo 4-17

Photo 4-18

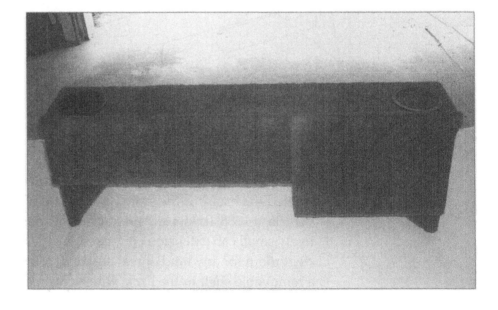

Photo 4-19

This has been a very good sounding system, and one I have not tired of over the years. The coaxes in the rear add significantly to the sense of space and tend to lift the soundstage much higher than the 4-in coaxes in the dash can do alone. They also fill in the midbass punch that the dash speakers lack—all in all, a good combination. To really polish the sound of this system, I added a 10-band graphic equalizer and tuned everything for flat response with the aid of the real-time analyzer. Of course, once that was done, I turned up the subwoofer volume to make for lively jazz and pop reproduction. At a sanctioned IASCA key event competition a number of years ago, this system recorded a sound pressure level of 127 dB. Not bad for two 8-in subs that cost $35 each.

That about wraps up this chapter on pickup trucks. There should be something here for just about any type.

Chapter 5

Rear Systems for Vans and Sport Utility Vehicles

In this chapter, we will look at some possible system arrangements for vans, sport utility vehicles, and Jeeps in particular. Due to the difficulties associated with the limited space available in a Jeep Wrangler, we will present a unique prototype bass cabinet to fit this specific application. We begin this chapter with a discussion on vans.

Vans

Figure 5-1 shows the two preferred options for installing a subwoofer system in a van—under the rear seat. Using this placement has a couple of benefits. First, it avoids taking up valuable space in the cargo storage area and, second, it provides a line of sight to the driver's seat. This aids considerably in projecting the bass sound pressure wave forward and minimizes the loss of detail and impact that would result from having it bounce around the rear cargo compartment. The bottom line is that this approach works surprisingly well considering the resonant nature of the van's space.

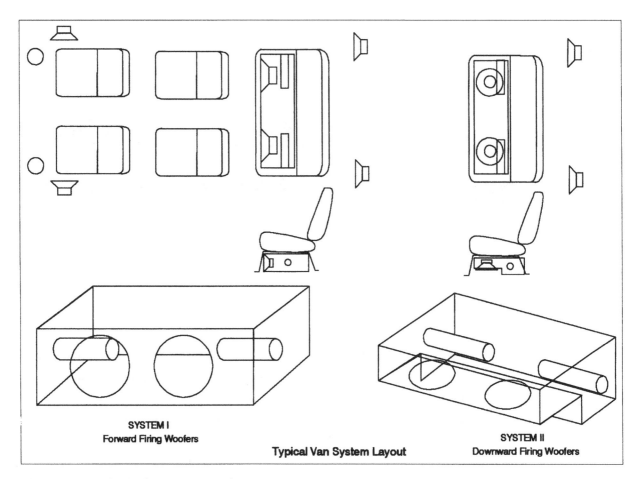

Figure 5-1   Typical van system layout.

Main speakers can be located in the dash, doors, walls, and rear roof pillars. Use as many as possible in order to achieve full, even sound distribution and overcome road and engine noise. If you install more than two pairs of speakers, be sure to make provisions for adjusting their relative volumes. For best results, use the same model series coax throughout the vehicle.

System I will fit in many full-sized vans where the distance between the seat bottom and floor can be quite large—as much as 12 in at the front. The bottom of the seat is usually not level and slopes downward from front to back. Therefore, to maximize the available space, you may want to consider a trapezoidal shape. When space is sufficient, the preferred woofer orientation is facing forward as indicated. Either 8- or 10-in versions are possible in some vans, and system designs can be any of those given in Figure 6-6 or 6-7. The recommended systems are any of the hatchback systems we built in Chapter 3 using the Madisound 8154s in a vented box or the 10207s in either a closed or vented box. You could even consider installing the full-range hatchback system if your current rear speakers are too distant to provide enough detail at the driver's position.

System II is designed to fit under the rear seat of a typical minivan. Because height is somewhat restricted, the woofers must be positioned to face down. In almost all cases, the box width is too narrow to aim the ports forward, and therefore they must be located in the sides. This works best when the seat legs are open-frame design; otherwise 2 in of clearance must be maintained between the port and seat frame. The box is stepped 2 in to make maximum use of available space and to provide a reflective surface to launch the subwoofer pressure wave. This concept works very well in practice. The recommended system in these compact applications is the vented version of the Madisound 8154.

If you can't fit anything under your rear seat, you can resort to the configurations shown in Figure 5-2. System III uses two tall narrow cabinets positioned to fire through the space between the rear seatback and cabin wall. These have the disadvantage of encroaching on the cargo space to some extent and are also quite visible, which means they really should be carpeted. In addition, you will have to provide some type of attachment to keep them in place. Designs for these systems are the single woofer units given in Figures 6-4 and 6-5.

When all else fails, you can resort to any of the hatchback units in Chapter 3 installed behind the rear seat facing up. These are represented as System IV. However, with cargo space at a premium during family trips, you will probably elect to remove the cabinet to make room for luggage.

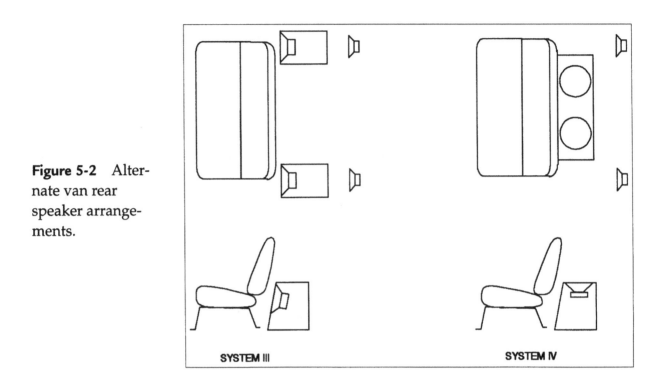

Figure 5-2 Alternate van rear speaker arrangements.

Sport Utility Vehicles

With few exceptions, sport utility vehicles don't have much, if any, space under the rear seat. There are usually contours in the floor pan that serve as legs to elevate the rear seat, and so it is bolted directly to the floor. For all of these applications a hatchback system behind the rear seat, with the drivers facing up, is a good solution. For best results, the cabinet should be sized so that the drivers are nearly level with the top of the rear seatback, as shown in Figure 5-3a. Since space is relatively abundant, you are free to choose any of the hatchback systems of Chapter 3 or 6.

Jeep Wrangler System

The open-air Jeep Wrangler is a special case of the sport utility vehicle, and is somewhat more challenging. Because there are no known, commercial subwoofer cabinets available for this vehicle, we decided to build one as a construction project.

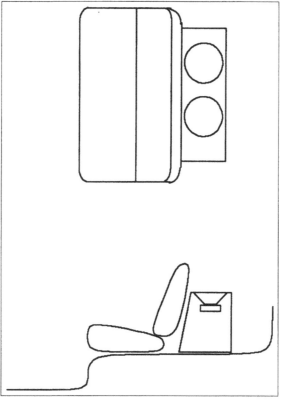

Figure 5-3a Typical sport-utility vehicle subwoofer arrangement.

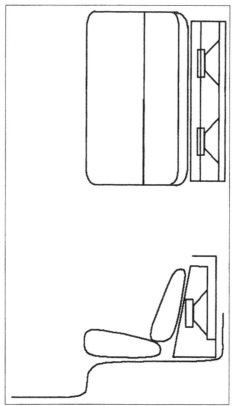

Figure 5-3b Jeep subwoofer arrangement.

To say that available space is minimal is an understatement. From an audio perspective, road noise will be considerably higher than in a car or pickup truck, and the Jeep's soft top is an unknown as far as acoustic properties are concerned. For all of these reasons, the sub-woofer cabinet we have designed for this application is intended to launch the woofers' pressure wave toward the front of the vehicle through a duct which "fires" over the top of the rear seatback. The duct has the added advantage of protecting the woofers from sunlight and moisture. On the negative side, under the right circumstances, it could scoop up dirt and debris when driving with the top down.

To be on the safe side, we designed this system for high output in free air. Therefore, we chose to use two Madisound 10207s in a 1.2-ft³ (net) sealed cabinet, which was tailor-made to fit the space behind the rear seat. Theoretically, this cabinet is a little undersized—1.5 to 1.7 ft³ would have been more ideal. However, as it turns out, the friction caused by the ducting lowers the system resonant frequency by about 25 percent and more than compensates for the reduced cabinet volume.

Getting Started

At the local Jeep dealer, we made some measurements of the cramped space behind the rear seat and transferred them to a poster-board template, which represents the enclosure cross section. We made a second trip to the dealer to verify that the template was accurate and made a few adjustments. Photos 5-1 and 5-2 show the template in place in the rear of the Jeep. As we did with the standard cab pickup truck design, we next made a full-scale drawing of the enclosure cross section on the template. Without this step, it would have been nearly impossible to build this cabinet. After all of the prework, the drawing shown in Figure 5-4 evolved. In this setup, the woofers fire toward the rear—the only orientation possible due to the narrow clearance between the top and the rear seatback. In addition, we had to make additional room in the front cabinet wall to accommodate the woofer magnet.

The sealed enclosure is constructed from our standard ¾-in high-density particle board. To maximize bass duct width, we made the

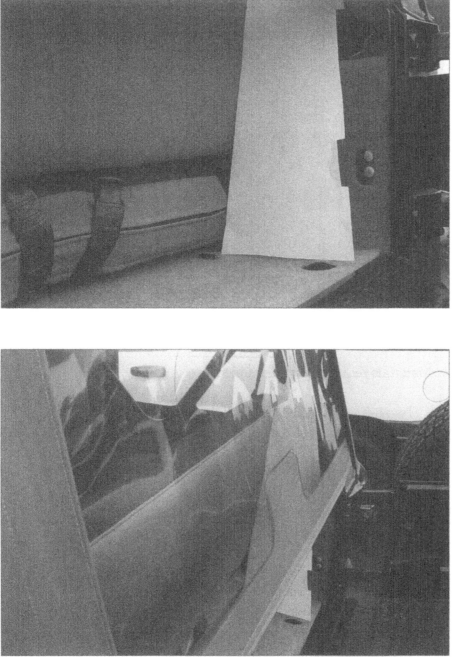

Photo 5-1

Photo 5-2

duct wall out of ⅝-in-thick material. Whether or not the additional ⅛ in achieved by doing this provides a measurable improvement is unknown. Directionally, it reduces velocity, which should therefore reduce distortion. For convenience, the top, bottom, and sides and duct can be made from the 11½-in-wide shelving.

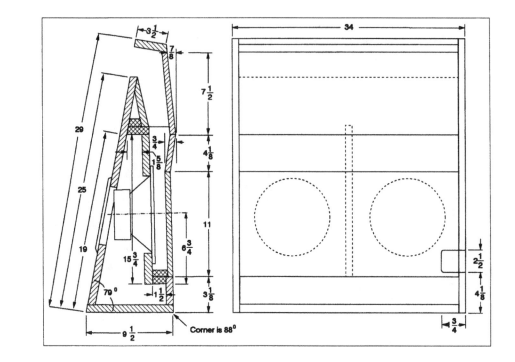

Figure 5-4 Jeep Wrangler bass cabinet.

Materials required for this project are:

- Two 8-ft sections of ¾-in particle board shelving
- One 8-ft section of ⅝-in particle board shelving
- One-half sheet of ¾-in high-density particle board (a full sheet would be required without the shelving above)
- Polyester fiberfill—2 lb
- Box of coarse-thread drywall screws, 1⅝ in long
- Yellow carpenter's glue
- Box of rope caulk
- 16 No. 8 panhead sheet-metal screws, 1 in long
- 8 No. 6 panhead sheet-metal screws, ¾ in long
- 2 Madisound 10207 10-in woofers with voice coils wired in parallel
- 2 terminal cups
- 4 ft of No. 16 or 18 speaker wire
- Solder

The parts used in this project are shown in Photo 3-35, with the exception of the grilles, which were eventually discarded.

Construction begins by taking measurements from the full-sized drawing on the template and transferring them to the corresponding 32½-in-wide pieces. Photo 5-3 shows the dry-fit pieces, which form the sealed cavity without the sides. Note the planned joint overlaps that were later trimmed off with the router.

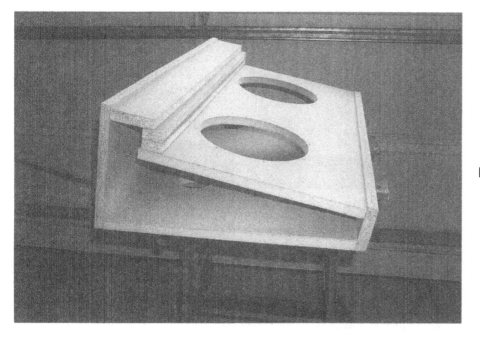

Photo 5-3

Photo 5-4 shows the interference between the woofer magnet and the front cabinet wall. This was anticipated from the layout on the scale drawing. Clearance was provided by gouging out this area with a Dremmel tool, as shown in Photo 5-5. However, initial testing disclosed that this was insufficient clearance for the magnet pole-piece vent and caused gross distortion at low frequencies, which resulted in the need to cut the holes in the wall shown in Photo 5-6. Note the unfortunate choice for the location of the terminal cup in the prototype. This prevented making the magnet relief hole even larger, as it should have. The terminal cup should be located well away from the woofer on future units. Photo 5-7 shows a ⅜-in-thick plywood plate glued and screwed over the magnet relief hole. This is not much thicker than the terminal plate and will not interfere with fitup in the back of the Jeep.

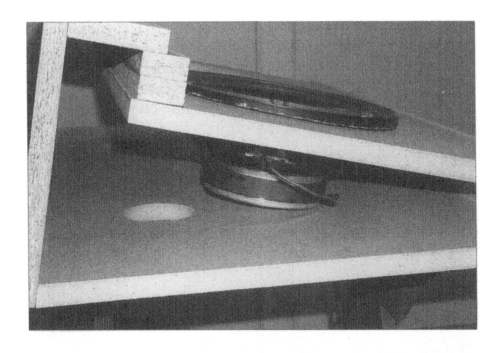

Photo 5-4

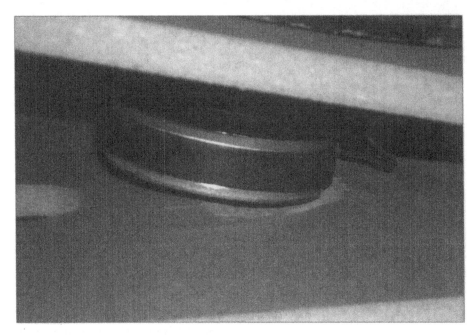

Photo 5-5

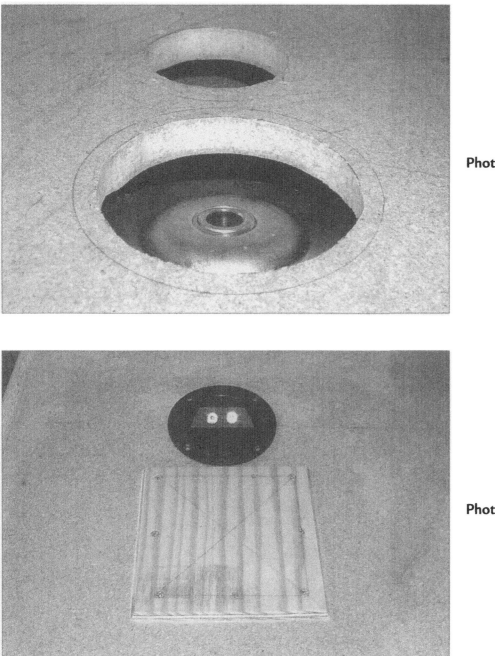

Photo 5-6

Photo 5-7

Photo 5-8 shows the box well along in the construction process. The clamps are holding a number of blocks being glued into place, which will eventually support the duct cover. A sloped extension will be added to the center support similar to the sides. In Photo 5-9, the completed box is shown without the drivers and access cover.

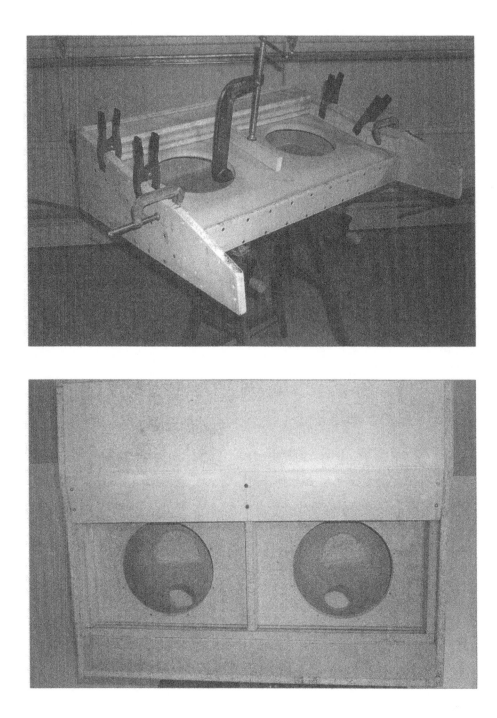

Photo 5-8

Photo 5-9

In Photo 5-10, the woofers were installed with grilles, the thought being that this would provide additional open-air protection. With all the space limitations in this system, these seemed to increase the low-frequency distortion problem and were eventually discarded. Also not shown in the photos is the 2 lb of polyester fiberfill stuffing that was later added after testing confirmed the need and an interior trapezoidal brace that ties the front and back panels together.

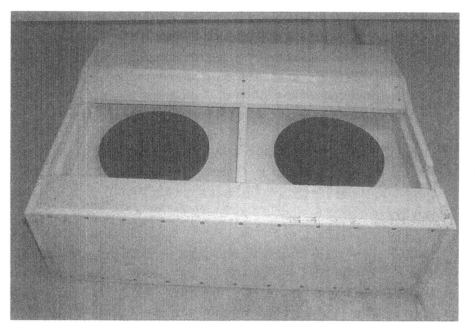

Photo 5-10

Photo 5-11 shows the completed box lying on its front with the woofer cover screwed into place. Note the notch routed into the lower right corner. It provides clearance for the Jeep's tailgate hinge.

Photo 5-11

Note: Before placing this cabinet in final use, the woofer cover must be sealed in with silicone caulk and allowed to cure for 24 hours. To ensure future accessibility, the caulk must be applied only to the edges (and two or three *small* spots on the center support) just prior to installing the cover. If there is a need to remove the cover, the caulk can be slit with a knife.

Further improvements could be made to the woofer access cover, which must be airtight and in this prototype was difficult to seal. Perhaps a piece of plexiglass and foam gasket would make a better alternative.

Photo 5-12 shows the Jeep box set up for initial testing. We'll come back to this a little later. In Photo 5-13, we're back at the Jeep dealer and confirm that the box fits as designed, and in Photo 5-14 (the moment of truth) the tailgate is shut without interference. Photo 5-15 shows how the box would look from the driver's seat. It appears that we could lower the height about an inch and improve the looks without loss of function.

Photo 5-12

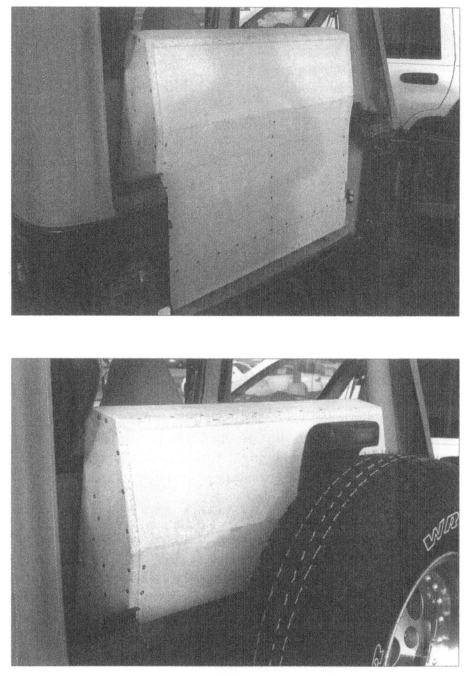

Photo 5-13

Photo 5-14

Photo 5-15

Testing

We measured the closed-box Thiele-Small parameters. Prior to cutting the magnet relief holes and installing stuffing, the closed-box system Q measured 1.05 with a resonance frequency of 49.2 Hz. After cutting the relief holes, removing the grilles, and adding stuffing, the Q dropped to 0.9 and the resonance frequency was reduced to 46.7 Hz—all in the right direction.

Figure 5-5a shows the predicted response for the Madisound 10207s in a closed-box system with a Q of 0.9. The solid line is the unboosted response, and appears quite similar to the measured response shown in Figure 5-5b. Note the rapid roll-off above 200 Hz caused by the duct.

We assembled a full-range test system, using the Jeep box as the subwoofer and listened to music in the garage. We adjusted the subwoofer filter by ear for best sound using minimum boost to minimize distortion and maximize headroom. The result is Figure 5-5c, which matches the predicted response well. The point here is that the Jeep box will be used in a mixed environment and needs to be tuned for the best compromise between extended bass and highest output levels for both free-air and in-car environments.

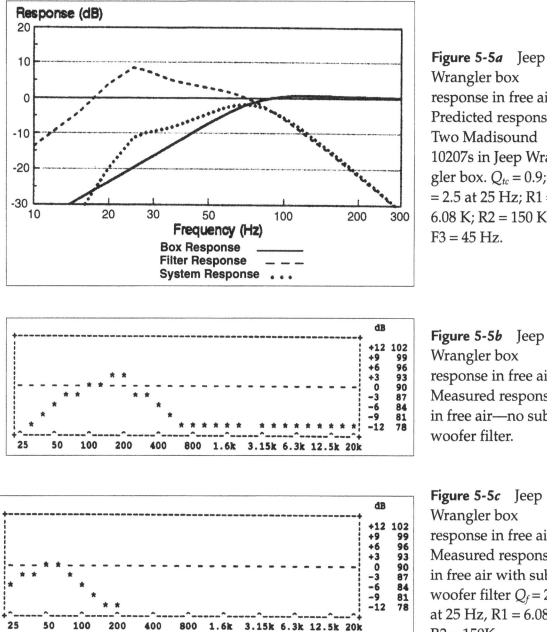

Figure 5-5a Jeep Wrangler box response in free air. Predicted response. Two Madisound 10207s in Jeep Wrangler box. $Q_{tc} = 0.9$; $Q_f = 2.5$ at 25 Hz; R1 = 6.08 K; R2 = 150 K; F3 = 45 Hz.

Figure 5-5b Jeep Wrangler box response in free air. Measured response in free air—no sub-woofer filter.

Figure 5-5c Jeep Wrangler box response in free air. Measured response in free air with sub-woofer filter $Q_f = 2.5$ at 25 Hz, R1 = 6.08K, R2 = 150K.

Since we were not able to test the Jeep box in an actual Jeep, we placed it in the back seat of a sedan and got the response shown in Figure 5-6. With the subwoofer filter set for no boost and a low-frequency cutoff of 30 Hz, we got another amazingly flat, picture-perfect curve. We should therefore expect reasonably good performance in the Jeep. Without low-frequency boost, the system easily handled 100 W rms of pink noise but generated a clean sound pressure level of only 105 dB. Efficiency is not a strength, as we might expect with this undersized box and the restrictions imposed by all the ductwork. A 200-W amp will get you another 3 dB, but that's about it.

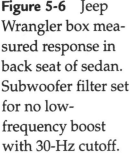

Figure 5-6 Jeep Wrangler box measured response in back seat of sedan. Subwoofer filter set for no low-frequency boost with 30-Hz cutoff.

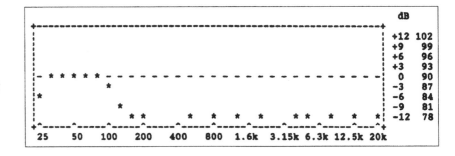

Finally, we took some acoustic output waveform measurements with a mike and oscilloscope, with the mike placed at a distance of 6 in from the outlet slot at a sound pressure level of 100 dB. These are shown in Photo 5-16. In Photo 5-16*a*, the woofers are exhibiting what appears to be overtravel in the forward direction at 30 Hz. Prior to cutting the magnet relief holes, this was orders of magnitude worse. Photo 5-16*b* shows the distortion is diminished at 40 Hz and appears gone at 45 in Photo 5-16*c*. Waveforms are totally sinusoidal from 45 Hz and up as shown in the rest of the photos. This box would have considerably higher output capabilities if it could be ported, and it would eliminate the low-frequency distortion problem.

Subjectively, the system sounded authoritative in the sedan. It was smooth and powerful and was capable of rattling all of the interior trimwork without signs of distortion. Attack was clean and the sound was definitely "tight."

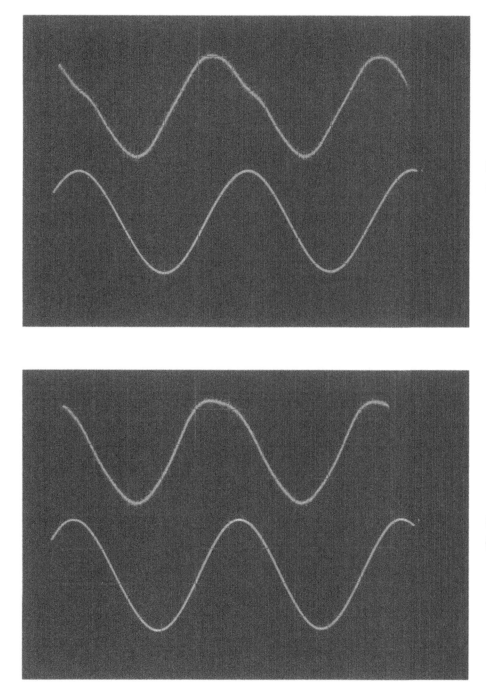

Photo 5-16*a*
(30 Hz)

Photo 5-16*b*
(40 Hz)

**Photo 5-16c
(45 Hz)**

**Photo 5-16d
(50 Hz)**

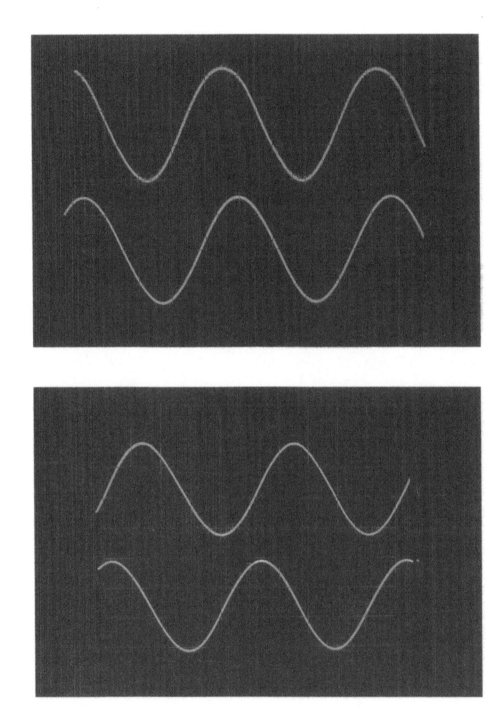

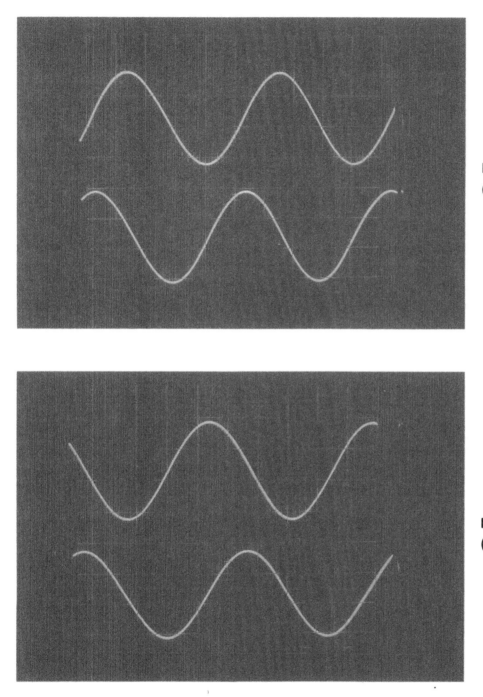

**Photo 5-16e
(100 Hz)**

**Photo 5-16f
(200 Hz)**

Two final points regarding this system. First, there is more work that could be done to increase system output. One way is by developing a port that exits at the top of the cabinet and dumps into the outlet slot. Another is to use drivers with higher excursion capabilities. Second, this was a very difficult project to construct and would have been extremely tedious to attempt without a table saw. However, if you have a Jeep Wrangler and just have to have some bass to go with it, this is a viable way to get it.

Chapter 6

More on Speaker Cabinet Construction

For those of you who are building your very first speaker system, you may be somewhat apprehensive as to what you're getting into. You may be seeing some new terminology for the first time, or you may have only limited carpentry skills. The math may even be somewhat vexing and, when you stop to think about it, most people just don't build their own speaker systems. But, from our numerous projects in the preceding chapters, you have probably seen by now that it's not that difficult.

The starting point is to understand that the three major components—drivers, cabinet, and filter—are all working together in a fairly precise relationship. The box volume and port dimensions are critically linked to the physical and electrical properties (referred to as Thiele-Small parameters) of the drivers. How they interact determines the filter settings which bring the whole system into focus. Any change to one or more of the parameters (like substituting drivers) without making the correct compensating changes to the others will result in a misaligned system and uneven bass response. Even if we execute the best designs available with great precision, the operating environment is still somewhat unpredictable and will play a huge role in determining the final system performance. Another advantage of using the filter is that it provides quite a bit of adjustment to compensate for many of these unknowns. As we saw in our construction projects in Chapters 2 though 5, the small volume in the typical vehicle passenger compartment provides quite a bit of low-frequency lift. Therefore, it takes much less boost to achieve 30- to 35-Hz response in a car than it does in free air.

It should be apparent by now that I rely heavily on my subwoofer filter to obtain better bass from small enclosures. Its operation and construction will be covered in detail in the next chapter. When a filter of this kind is applied to a ported box system, the concept is referred to as a "sixth-order Thiele-Small alignment" or when used in a sealed box application it becomes an "electronically assisted fourth-order alignment." The main features are extended low-bass response and built-in rumble suppression. The latter is of particular benefit with ported systems because it eliminates a common problem—driver cone flapping at very low ("infrasonic") frequencies—and aids in making the bass sound "tight."

While I personally would not build a sound system without employing my subwoofer filter, others may find it unnecessary or even undesirable. One of the objectives of this chapter is to explore alternative alignments which could be implemented either with or without my filter.

In every aspect of my design philosophy, I have attempted to provide low-cost and high-quality solutions to the auto sound problem. In this same context, I have selected woofers supplied by Madisound in Madison, Wisconsin, because they represent good value in price and performance. Even more important is that these drivers also permit designing relatively small enclosure volumes and are particularly well suited to bass reflex applications.

Madisound woofers may not be suitable for everyone. While they will play loudly enough for normal listening, they clearly don't have the power-handling and high excursion capabilities found in some drivers designed exclusively for auto sound. For these reasons, we'll look at some alignments based on drivers from two of the major auto sound manufacturers, Kicker and MTX. In general, their products are more rugged and are capable of generating higher sound pressure levels, making them more appropriate for auto sound competition events, sometimes referred to "sound offs" or "crank 'em ups."

Construction Rules

When building any speaker cabinet, the following general rules must be adhered to to assure that the system will perform properly.

1. Speaker cabinets must be airtight. Any leakage will cause misalignment of the system and result in irregular bass response.

2. The cabinets must be nonresonant. In other words, the walls and structure must be rigid and free of rattles or buzzes. A good test for this is to listen to the sound made by rapping on the walls with your knuckles. If the wall is stiff, the sound will be sharp, similar to the sound of tapping on a brick. If the sound is dull and hollow, the wall is probably too flexible and needs additional bracing.

3. Actual box dimensions are not critical and can be adjusted to suit your specific application as long as internal box volumes and port dimensions are maintained. Therefore, remember to take the volume of any internal bracing into account when sizing your cabinets.

4. In any ported design, you must maintain at least one duct diameter clearance between the end of the duct and the opposing cabinet wall to prevent misalignment of the system. Any restriction in airflow tends to make the effective length of the duct longer. For best results locate the end of the duct in as much free space as possible.

5. To significantly reduce "wind noise," which is a form of distortion, be sure to round over the edges of the duct with a file (or router) and then sand them smooth.

6. Ported cabinets should be lined with acoustic fiberglass like that sold at Radio Shack. The material should cover all interior walls except the one the drivers are on. To work properly and avoid problems with obstructing the port, it must be securely stapled down. Take the insulation thickness into account when figuring duct clearance above.

7. High-density particle board is the preferred cabinet material. It comes in various thicknesses, but most cabinets can be successfully built from ¾-in-thick material.

8. When building cabinets from particle board, glued and screwed butt joints are a convenient and strong construction method. Coarse-thread drywall screws appear to be ideal for

this purpose. There are a total of three drilling operations that must be performed, so using one of the new quick-change drill attachments (or two drills) will usually speed assembly considerably.

9. For purely acoustic reasons, drivers should be mounted to the outside face of the baffle. This also makes removal a snap if one ever fails and you have to replace it. It also provides a relatively easy way to have access back inside the cabinet if you decide to make some internal modifications.

10. Place two or more coils of rope caulk on the back of each driver's mounting flange prior to installing it in the cabinet. This makes for an airtight seal and one that can be removed with relative ease in the future. The number of coils of caulk needed is determined by the width of the driver flange. On small drivers, perhaps only one is required; while on large woofers, as many as three will be needed.

11. When metal grilles are used which employ a mounting ring that is sandwiched between the driver and cabinet wall, two rope caulk gaskets are required per driver. One goes between the back of the driver and the ring, the other between the ring and cabinet.

12. Do **not** rely on carpet as a driver gasket. It leaks!

13. Drivers should be screwed into the baffle board with No. 8 panhead sheet-metal screws. Good lengths for these screws are 1 or 1¼ in. Drywall screws are an acceptable substitute for use on some woofers, provided that they seat properly into the mounting flange. For mounting tweeters, you will typically need No. 6 oval or flathead sheet-metal screws.

Tools

As seen in previous chapters, you will need the following tools:

1. Electric circular saw ("Skil saw") with sharp, carbide-tipped blade
2. Saber saw
3. Variable-speed, reversible electric drill with drill bits, countersink, and screwdriver bits
4. Clamp-on straightedge

5. Carpenter's square of good quality
6. Carpenter's steel tape, ¾ or 1 in wide
7. Caulking gun
8. Screwdrivers
9. Wood rasp
10. Compass

Procedure

Make a Cutting Diagram

Figure 6-1 shows typical construction details. Start by making a rough scale drawing of the cabinet. Then sketch and dimension each piece to minimize cutting errors. Use these sketches to make a "cutting diagram" like the one shown in Figure 6-2. While all this may not seem necessary, there are some good reasons for doing it this way. First, it minimizes waste. (You could even avoid buying an extra sheet of particle board in some cases.) And, it has the potential to improve accuracy, since pieces with common dimensions are cut to length at the same time.

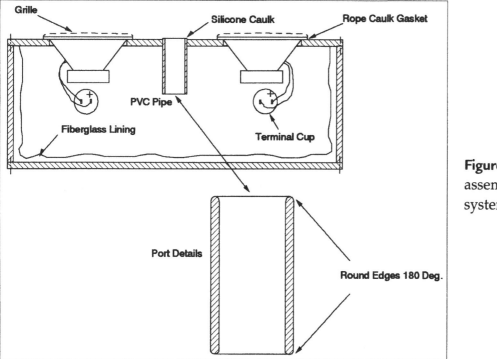

Figure 6-1 Typical assembled ported system.

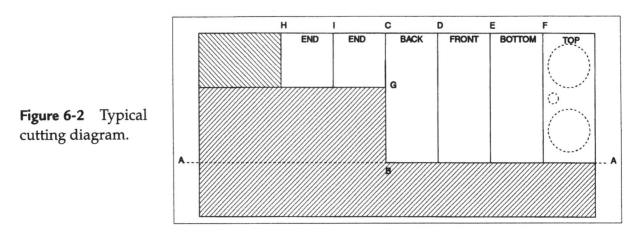

Figure 6-2 Typical cutting diagram.

Cut Out the Pieces

Referring to the cutting diagram, the first step would be to rip along line A. You have two options here. You could rip the entire 8-ft length off the board or you could stop at some point B to conserve material and create a useful piece of scrap. Next, make cuts C, D, E, and F, and the four big pieces are cut to size. Three more cuts—G, H, and I—complete the process. It's that simple.

Unless you have a giant table saw like Norm Abrams (ETV's New Yankee Workshop), the easiest way to make these cuts is with a clamp-on straightedge and **sharp** circular saw, as was shown in Chapter 2. Due to the abrasiveness of the particle board, a carbide-tipped saw blade is almost mandatory. The straightedge is a $25 item that will make the difference between a merely acceptable job and one you will be proud of. It will take a little practice to become proficient with it, so try it out on some scrap before you make that first big cut. During the actual cutting process, try not to bear too hard against the straightedge to avoid making it slip.

After all of the pieces are cut to size, lay out everything—screw holes, mounting holes for drivers, terminal cups, and port (if any). Draw everything in with as much precision as possible. Try your best to avoid oversizing driver mounting holes, as it will cause real problems later on at final assembly.

After cutting all the holes out, test-fit each of the components. A wood rasp with a curved side really comes in handy for doing touch-up work to the holes.

Drill the Pilot Holes for Mounting the Components

It's now time to drill all of the pilot screw holes for mounting the drivers and terminal cups. If you measured and cut carefully, your components just barely fit into their holes. One at a time, place each component in its corresponding hole and, after centering it carefully, trace all mounting screw holes with a sharp pencil. Then remove the part and examine the marks. Verify that they are evenly spaced from the edges of the holes. If any appear to be off center, go back and retrace. When all is well, lightly center-punch each hole and drill all of the pilot holes for snug screw fit. A $\frac{3}{32}$- or $\frac{7}{64}$-in drill bit works pretty well for No. 8 sheet-metal screws.

Drill the Cabinet Assembly Screw Holes

Lay out the screw lines $\frac{3}{8}$-in (or one-half the board thickness) from the edge of the boards where the screws will be located; then mark and center-punch the screw centers on the lines. If you are planning for any overlap in your joints, add this amount to the edge spacing above. Space holes evenly along each edge with no more than 3 in between centers. Also, do not locate screws closer than $1\frac{1}{4}$ in from the ends to avoid splitting the pieces you will be attaching to. Drill through the marked holes with a $\frac{5}{32}$-in drill bit and then countersink. As shown in Photo 2-27, an oversized drill outfitted with a secure stop makes countersinking a breeze. A $\frac{3}{8}$ drill works well for No. 6 drywall screws.

Dry-Fit the Pieces

With all of the "through holes" drilled and countersunk, dry-fit all of the pieces together and attach each piece with two or three dry wall screws. Drill pilot holes in the receiving pieces with a $\frac{3}{32}$-in drill. This will take more time than you might think to do correctly as there are never enough hands or clamps to go around. The corner clamps shown in Photo 2-28 are invaluable for the dry-fitting operation.

When positioning the boards, remember it is always better to have a **slight** overlap at the edges than underlap. The reason is that you can remove the overlap with the rasp, router, or belt sander and make the job look very professional. Underlaps are just plain ugly.

Assemble the Cabinet

Assemble the cabinets carefully with plenty of yellow carpenter's glue. Wipe off excess with a damp cloth as you go. You will need to rinse the cloth often, so place a bucket of warm water nearby. Allow the assembled cabinet to dry overnight or 2 to 3 hours in direct, hot sunlight. If you are doing this project during the winter, it's a good idea to place the cabinet in a heated room to cure the glue.

After the initial glue is thoroughly dry, you can seal minor joint gaps by flowing generous amounts of additional glue into interior joint lines. Let this stand again until thoroughly dry. Inspect the interior carefully and verify that the cabinet is completely sealed. If there are still visible gaps, seal them with silicone caulk from the inside.

Round the Corners

If you have one, now is the time to get out the router and remove any joint overlaps using a flush cutting bit like the one we showed in Photo 2-35. A belt sander will do this job fairly well also. For a really professional look, round all the cabinet edges over with a router and ⅜-in roundover bit. Lightly sand the routed edges smooth by hand. If you don't have access to the power tools, a decent job can still be done with a wood rasp, sanding block, and elbow grease. It just takes a little more time and patience.

Paint the Cabinet

Prior to painting, fill in any holes or other glitches with water-based wood putty. If you want your screw holes to be invisible, remove all of the drywall screws prior to filling. It will probably take two applications to completely hide the holes, similar to finishing drywall. For this job, I prefer premixed Elmer's or "Durham's rock hard wood putty," which comes in powder form. The advantage with the dry material is that you can mix a fresh batch each time you use it to the consistency you prefer. When mixed with a little extra water, it can be used very effectively to fill in the pores in the edges of the particle board. Allow it to dry overnight and then sand smooth with 100-grit paper.

For final finish, I prefer satin or semigloss black. If you want a professional look, apply several coats and sand off any runs between coats. The best paint I have found for this comes from Wal-Mart and happens to also be very inexpensive—about $1.50 per can. It will usually take about two cans to provide enough paint for multiple coats.

While a good paint job is an acceptable way to finish a speaker cabinet, in many installations, a more attractive alternative is carpeting. However, since it's the last step in the process, we'll discuss it a little later.

Install the Wiring

The first step in the final assembly is to install the wiring and terminal cups. Obtain a roll of 16- or 18-gage speaker wire (zip cord with one lead marked). Cut off about 3 ft and solder to each terminal cup. Maintain polarity by connecting the striped wire to the red (positive) terminal. Feed the wire into the box and install the terminal cups with some ¾-in-long No. 6 panhead sheet-metal screws. Most terminal cups come with a rubber gasket, which should provide a sufficient seal. If none is provided, apply a bead of silicone caulk to the edge of the hole prior to installing the cup.

Staple in the Fiberglass

Next, line the box with acoustic fiberglass like the kind sold at Radio Shack. Do this neatly, as it is lining, not stuffing. Cut the pieces to approximate size and staple in place. This material is best cut with a pair of scissors. Cut a slit in the fiberglass and feed the speaker wire through it and out the speaker mounting holes.

If you are building a sealed box design, fiberglass is not used. Instead, lightly stuff the box with polyester pillow fiber—about 1 to 1.5 lb per ft^3.

Install the Port

Now install the port (if there is one), which is usually a piece of PVC pipe. In many cases you will have to buy a full 10-ft "stick" to get your small piece. (You can save a few bucks by buying it from Lowes, which also sells it in 5-ft lengths.) Cut it to the specified length with a hacksaw; then square up the ends and round over the edges with a file. If you have access to a table saw, it does a superior job of cutting PVC pipe square as shown in Photo 2-31. Spray paint the inside surface and outer edge with flat black paint and let dry. Once we start applying silicone caulk, paint will never stick there well again.

The pipe should fit slightly snug in the baffle board port for easiest assembly. Apply a generous bead of caulk in a ring around the outside circumference of the pipe about 1 in from the outlet end. Insert the opposite end of the pipe into the hole in the baffle board. Slowly push the pipe in until the caulk ring contacts the board. Then use a twisting motion to fully distribute the caulk as you push the duct flush with the surface of the board. If you plan to install carpet, leave about ¼ in of pipe protruding. This should end up about even with the surface of the carpet and will look much nicer than when the pipe is simply flush with the cabinet. In either case, allow the silicone to cure until firm before using. Direct, hot sunlight will speed up the process considerably.

Install the Grilles

We are now ready to mount the drivers. Begin by installing the metal grille mounting rings on the rear of the drivers. The ones sold by Madisound have only four of the eight holes predrilled in the 8-in size, so you will need to mark and drill the other four before proceeding. **Be extremely careful to not get any metal shavings near the driver magnets while doing this.** Make a gasket of rope caulk that completely covers the rear of the driver mounting flange and press the grille ring into place. Be sure all the screw holes are aligned.

Connect the Wiring

Next solder the wires to the speaker terminals. **Do not apply excessive heat or the driver voice coils will be damaged!** Remember to maintain polarity by connecting the "plus" or red terminal on the external connections to the driver terminal with the red dot.

Install the Drivers

Make another gasket of rope caulk wide enough to cover the rear of the grille ring and press it into place. Press all eight sheet-metal screws through both layers of caulk and set the driver/grille ring assembly in place. Start each screw by hand into its pilot hole and then tighten down evenly. **Be extremely careful while tightening these screws. One slip, and you have a damaged speaker!** Trim off the excess caulk after it squeezes out; then press the grilles on.

Install the Carpet

If you plan to install carpet, now's the time to do it. You have two options. For the best appearance, you could obtain a piece that matches your car's interior from an auto upholstery shop. The problem with this is that this type of carpeting will be stiffer and more difficult to work with. If you can live with a neutral color like gray or black, it's better to use carpet specifically designed for speaker cabinets which is available at your local auto sound installer. There are lots of ways to do this, but the easiest I've found is to use silicone caulk instead of the usual spray contact cement. The reason is that the procedure I'm about to give you is much more forgiving. Also, conventional wisdom says it's OK to carpet the system before installing all of the components and then use the carpet as if it were gasket material. I strongly disagree with this approach, as carpet makes a very poor gasket.

For our carpeting to look "professional," the one thing we want to avoid is having any visible seams, and we certainly don't want any of them located on cabinet edges. At first this may seem impossible, but it's not. Referring to Figure 6-3a, we begin by wrapping the completed system in carpet as if it were a package. We will locate the seam in the middle of the back of the enclosure where it is seen very little and also where it will not be damaged by sliding the cabinet along the floor.

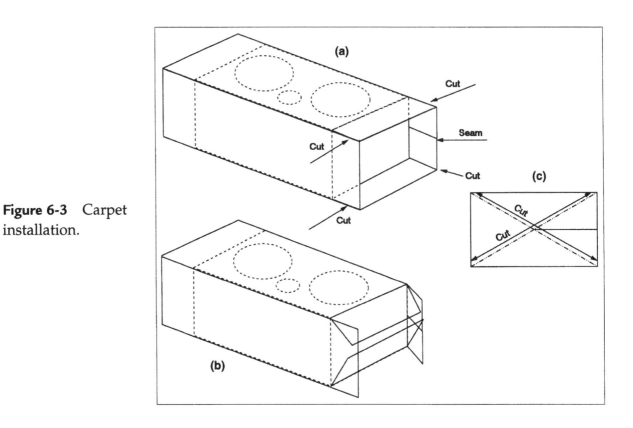

Figure 6-3 Carpet installation.

I find that a kitchen table that you can walk all the way around is a convenient place to apply carpeting. Begin by laying the carpet face down on the table and then placing the completed speaker system on top as shown in Photo 6-1. Staple one edge of what will be the rear seam to the box (Photo 6-2); then bring the rest of the material up and over to see how much excess material is available. Leave 2 or 3 in of overlap and cut off the excess. With the original staples still in place, unwrap the enclosure and roll it over onto its rear surface (where the seam will be). Apply silicone caulk to the bottom, front, and top surfaces of the enclosure as shown in Photo 6-3. Place the caulk in a big X pattern on each panel and along each edge. On the top, outline the speaker grilles and port (which should be protruding slightly) with beads of caulk. Use a putty knife to spread it around if you need to.

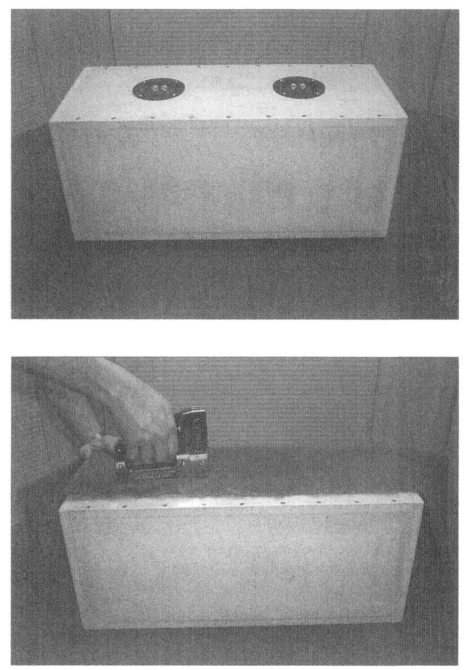

Photo 6-1

Photo 6-2

Photo 6-3

Rewrap the enclosure by rolling it over until the seam area is on top. As you do this, secure the carpet on all three glued surfaces with plenty of staples. Now remove the original seam staples and apply the caulk as before. Be sure to put generous beads adjacent to where the seam will be located. Fold the original flap back into place, which should still be located in the center of the panel, and staple it along the edge. Add other staples to make it secure. Now for the moment of truth. Fold the last flap back over on top of the first. Allow a slight overlap and staple it securely in place as shown in Photo 6-4. Using a large, **sharp** knife, remove the excess material by cutting through **both** flaps at the same time as shown in Photo 6-5. Touch up your silicone application in the seam area and secure the last flap with plenty of staples. Done properly, the main seam should now be almost invisible.

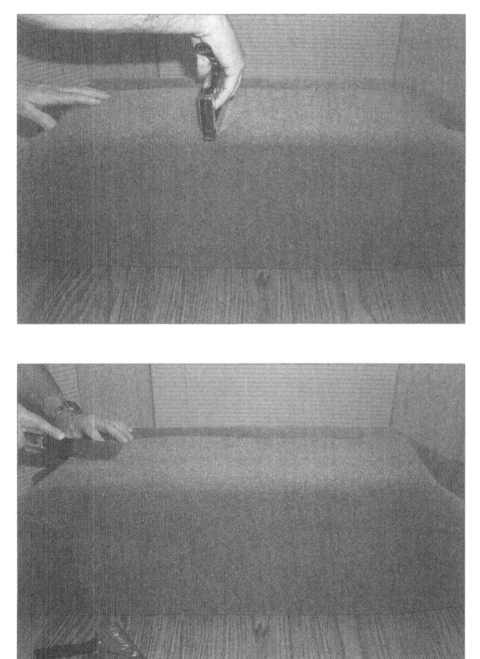

Photo 6-4

Photo 6-5

Roll the enclosure over so that the drivers are now facing up. Using a razor knife (like the kind used for wallpaper), trace around the grilles and port (if there is one) as shown in Photos 6-6, 6-7, and 6-8. Remove the scrap pieces. If you need to, lift the carpet and touch up the caulk as shown in Photo 6-9. Then press it back in place and add more staples as necessary to secure it as shown in Photo 6-10.

Photo 6-6

Photo 6-7

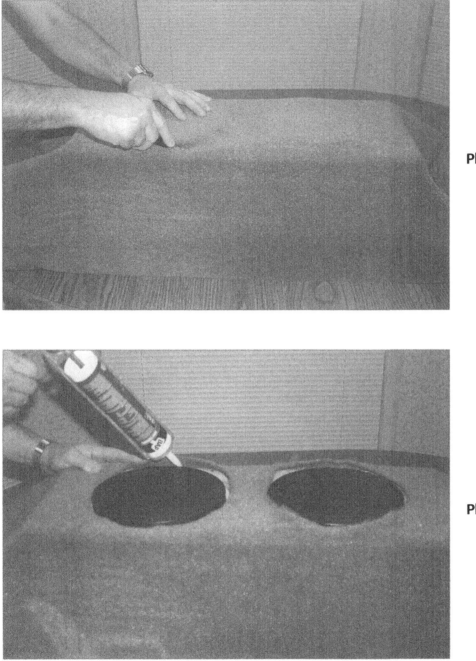

Photo 6-8

Photo 6-9

Photo 6-10

Referring back to Figure 6-3*a*, make the four end flap cuts as indicated and shown in Photo 6-11. Fold the flaps over on top of each other, again as if you were wrapping a package as shown in Figure 6-3*b*. Staple them down to secure them in place. Resharpen your knife and cut through all of the layers diagonally as indicated in Figure 6-3*c* and shown in Photo 6-12. Remove the scrap and verify that all the triangular flaps fit together neatly to cover the end as shown in Photo 6-13. If everything looks OK, apply caulk to all seam areas and press the flaps into place. Secure them with more staples. Repeat the process for the opposite end.

Let the cabinet stand overnight. The next day, remove all the staples with a screwdriver, fluff up the carpet, and enjoy your handiwork. The finished product is shown in Photos 6-14, 6-15, and 6-16.

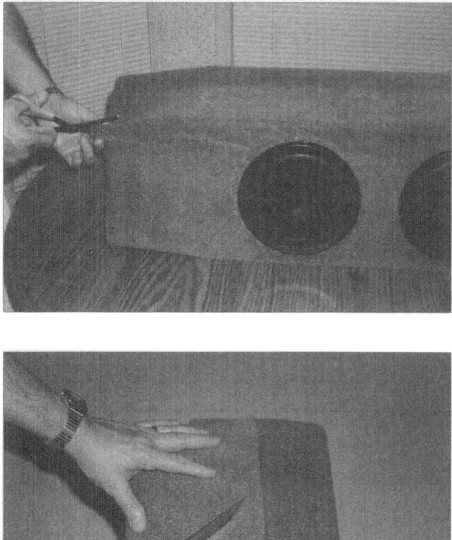

Photo 6-11

Photo 6-12

Photo 6-13

Photo 6-14

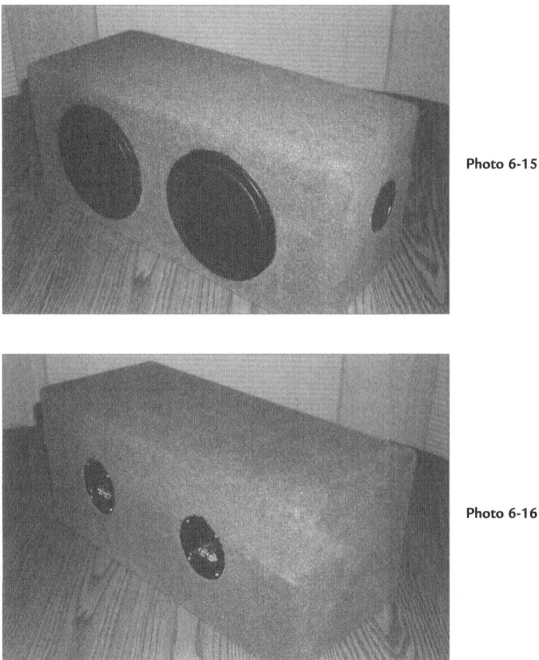

Photo 6-15

Photo 6-16

Other Possibilities

Now that we've developed a good understanding of the principles of speaker enclosure construction, let's explore a few more cabinet design possibilities.

Figure 6-4 is a compact closed box for a single bass driver, which can be used singly or in multiples. In general, closed systems can be made smaller than ported systems. However, the trade-off is low efficiency, which translates into a need for more amplifier power. And because the driver does all the work in a closed system, power handling and maximum sound pressure levels will be lower than for an equivalent ported system. On the plus side, closed-box alignments are much more forgiving than ported boxes, so a wider variety of drivers can be used.

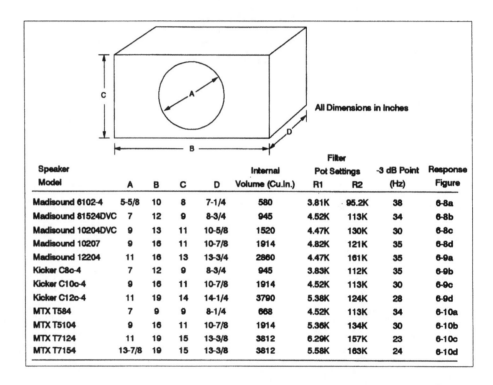

Figure 6-4 Sealed-box alignments for various drivers.

Speaker Model	A	B	C	D	Internal Volume (Cu.In.)	Filter Pot Settings R1	R2	-3 dB Point (Hz)	Response Figure
Madisound 6102-4	5-5/8	10	8	7-1/4	580	3.81K	95.2K	38	6-8a
Madisound 81524DVC	7	12	9	8-3/4	945	4.52K	113K	34	6-8b
Madisound 10204DVC	9	13	11	10-5/8	1520	4.47K	130K	30	6-8c
Madisound 10207	9	16	11	10-7/8	1914	4.82K	121K	35	6-8d
Madisound 12204	11	16	13	13-3/4	2860	4.47K	161K	35	6-9a
Kicker C8c-4	7	12	9	8-3/4	945	3.83K	112K	35	6-9b
Kicker C10c-4	9	16	11	10-7/8	1914	4.52K	113K	30	6-9c
Kicker C12c-4	11	19	14	14-1/4	3790	5.38K	124K	28	6-9d
MTX T584	7	9	9	8-1/4	668	4.52K	113K	34	6-10a
MTX T5104	9	16	11	10-7/8	1914	5.36K	134K	30	6-10b
MTX T7124	11	19	15	13-3/8	3812	6.29K	157K	23	6-10c
MTX T7154	13-7/8	19	15	13-3/8	3812	5.58K	163K	24	6-10d

Two of these cabinets with 8-in drivers should be considered the minimum. When space constraints dictate, four 6-in drivers can be used as an acceptable, although less desirable, alternative.

Some comments about the specific closed-box alignments chosen in Figure 6-4:

- The Madisound 6102-4s have decent parameters for 6-in drivers and can fit into something not much bigger than a shoebox—their only real virtue as a subwoofer. With the addition of some stuffing, these will easily align at the ideal closed-box Q of 0.7 in a 0.3-ft³ enclosure. But, this system is too small to be taken seriously, especially since the drivers have only a 3.5-mm excursion limit. Without low-frequency boost from the filter, this system is down 3 dB in free air at only 65 Hz, which is not real bass. In a car this could improve to 45 to 50 Hz, but clearly these drivers will tolerate little if any low-frequency boost.

- The Madisound 81524DVCs have a 5-mm peak excursion, which is good for an 8-in driver, and the dual voice coils provide added flexibility in application. By using only one of the voice coils, the Q can be raised to 0.6, making them a natural for free-air applications. When used in a 0.5-ft³ enclosure with both voice coils driven, they have a closed-box Q of 0.6 without stuffing, which in this case should be added only sparingly. Unboosted, they are down 3 dB at 55 Hz and should easily reach about 40 Hz in-car. A slight boost will take them to 35 Hz.

- The Madisound 10204DVCs have a decent excursion limit of 6 mm and also fit well into a small box. With a box size of only 0.8 ft³, they have a closed-box Q of 0.75, which will easily drop to 0.7 or less with some stuffing. Unboosted, they should be down 3 dB in free air at about 55 Hz and make 35 Hz in-car. A modest boost in-car will easily generate flat response to 30 Hz. This system is a good candidate to try without the filter.

- The main difference between the Madisound 10207s and 10204DVCs is voice coil impedance—two 4-ohm voice coils in the 10204 and two 7-ohm coils in 10207. The choice you make between the two would be based primarily on the type of amplifier hookup you're trying to achieve. However, in this particular example, we chose to use only one of its two voice coils, which results in a closed-box Q of 1.2. As we saw in our hatchback project in Chapter 4, this system generated a very extended in-car bass response with no boost, making it a good candidate to use without the filter.

- The Madisound 12204 needs 3 ft³ to achieve the more ideal Q of 0.75. In the interests of conserving space, we cut this in half and still got a respectable Q of 0.94. Unboosted, this system should achieve 43 or 44 Hz in free air with a stuffed enclosure. In a car, 30 to 35 Hz should be a cinch. This would make a good candidate for an SUV, where a cubical enclosure shape is acceptable.

- The Kicker C8c-4 is considered to be a "competition" model, but it has a rated excursion limit of only 5.17 mm, making it not much different from the Madisound 81524DVC. In a 0.5-ft^3 enclosure, it has a closed-box Q of 0.8, which is fine. Unboosted, it will be down 3 dB at 60 Hz in free air, which translates to about 45 Hz in-car. With a little boost, it will easily reach 35 to 40 Hz in-car.

- The Kicker C10c-4 aligned at a nearly ideal closed-box Q of 0.76 in a 1-ft^3 box, which will easily hit the ideal 0.7 with some stuffing. With an excursion limit of 7.9 mm, this driver should pack quite a wallop. Predicted –3-dB point in free air is about 49 Hz in free air with no boost. In-car, expect about 35 Hz. A modest boost with this driver will easily allow it to achieve flat response to 30 Hz. A pair of these in a hatchback coupled to a high-powered amplifier would be pretty brutal. This is also an excellent candidate to try without the filter.

- The Kicker C12c-4 aligned pretty well in a 2-ft^3 closed box with a Q of 0.9. Bass extension in free air was excellent, with a –3-dB point of 36 Hz in free air. Expect 25 to 30 Hz in-car with no boost.

- The MTX T584 has a huge (for an 8-in driver) excursion limit of 9.1 mm. And it aligns at a closed-box Q of 0.62 in a shoebox-sized 0.35 ft^3 enclosure. But, unboosted, it is down 3 dB at a high 70 Hz. In-car, this may only improve to 50 or 55 Hz. In this configuration it needs boost to realize its full potential. This is a brute-force, low-efficiency alignment, but if you have plenty of amplifier power (150 W minimum) and the need to cram a subwoofer into a very small space, this can work. With enough power input, it will also be loud. This will make a killer small system with the filter.

- The MTX T5104 also has a very high excursion limit of 9.4 mm. In a 1-ft^3 enclosure, it aligns with a closed-box Q of 0.84, which will easily be under 0.8 with some stuffing. It's down 3 dB at about 45 Hz in free air, so it should easily reach 30 Hz without boost in-car. This is a potent driver, apparently well suited for competition sound pressure levels.

- The MTX T7124 is built like a tank, as evidenced by its cast frame and 102-oz magnet. It has a huge excursion limit of 12.7 mm (or ½ in). Cramming all this firepower into a 2-ft^3 box while maintaining a reasonable closed-box Q of 0.88 means that efficiency is low. You will need 2 to 300 W per channel to do this driver justice, and the enclosure really should be 1 in thick with plenty of bracing. Unboosted, it's down 3 dB at about 35 Hz in free air, which translates to approximately 25 Hz in-car. I wouldn't stand too close to one of these at a sound-off.

- The MTX T7154 is a bigger version of the T7124. Whatever comments are made about the 12-in version are doubly appropriate here, as it appears to have the same voice coil and magnet structure. The good news, though, is that this 15-in driver aligns about as well in a 2-ft^3 enclosure as other 12-in drivers with a closed-box Q of 0.83. Unboosted, it has a –3-dB point of about 35 Hz, same as the T7124. With a little boost, it should be easy to generate 20 Hz in-car. With the added cone area I would rate this driver as "dangerous."

Figure 6-5 provides the vented versions of Figure 6-4. Being vented makes them able to handle more power in the critical low bass frequencies where the port is doing most of the work. In a Thiele-Small sixth-order alignment, the boost is added at a frequency which is near the box frequency. At box frequency, the driver cone moves very little in a well-constructed system. Therefore, the energy supplied by the boost has much less tendency to overdrive the woofers. Using an underdamped high-pass filter to supply the boost means that the rumble-suppression function is already built in, making a ported system sound much tighter than the simple boost circuit found onboard power amplifiers.

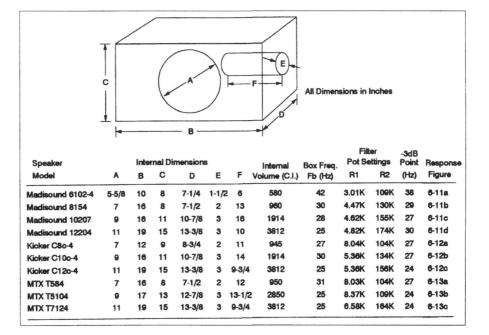

Figure 6-5 Vented-box alignments for various drivers.

Speaker Model	Internal Dimensions						Internal Volume (C.I.)	Box Freq. Fb (Hz)	Filter Pot Settings		-3dB Point (Hz)	Response Figure
	A	B	C	D	E	F			R1	R2		
Madisound 6102-4	5-5/8	10	8	7-1/4	1-1/2	6	580	42	3.01K	109K	38	6-11a
Madisound 8154	7	16	8	7-1/2	2	13	960	30	4.47K	130K	29	6-11b
Madisound 10207	9	16	11	10-7/8	3	16	1914	28	4.62K	155K	27	6-11c
Madisound 12204	11	19	15	13-3/8	3	10	3812	25	4.82K	174K	30	6-11d
Kicker C8c-4	7	12	9	8-3/4	2	11	945	27	8.04K	104K	27	6-12a
Kicker C10c-4	9	16	11	10-7/8	3	14	1914	30	5.36K	134K	27	6-12b
Kicker C12c-4	11	19	15	13-3/8	3	9-3/4	3812	25	5.36K	156K	24	6-12c
MTX T584	7	16	8	7-1/2	2	12	950	31	8.03K	104K	27	6-13a
MTX T5104	9	17	13	12-7/8	3	13-1/2	2850	25	8.37K	109K	24	6-13b
MTX T7124	11	19	15	13-3/8	3	9-3/4	3812	25	6.58K	164K	24	6-13c

The following comments concern the vented box alignments given in Figure 6-5:

- The Madisound 6102-4s will have less tendency to bottom out in a vented system due to the reasons given above. Still, it should only be used when nothing else will fit, as the "sound" will probably be "dry." It does align fairly well in the 0.3 ft³ vented box even though it is somewhat undersized. Expect it to reach about 40 Hz in-car with boost.
- The Madisound 8154s sound terrific in a 0.5-ft³ vented box, have good power handling, and can easily hit 30 Hz in-car with boost. Without boost, expect about 50 Hz in free air and about 35 Hz in-car.
- The Madisound 10207s align well in a 1-ft³ vented box. With their dual 100-W voice coils and 6-mm excursion limit, they can handle tremendous power. In-car, you expect −3-dB points in the low 20s with moderate boost. Without boost, the alignment shown will be down 3 dB at 45 Hz in free air or about 35 in-car.
- The Madisound 12204 does not fit well in a moderate-sized box. In the alignment shown, the 2-ft³ box is undersized quite a bit and response ripple is excessive. Without boost, the system has a −3-dB point of 40 Hz in free air, which translates to about 30 Hz in-car. With boost, you can expect 25 Hz in-car.
- The Kicker C8c-4 aligns OK in a 0.5-ft³ vented box. Without boost, expect it to be down 3 dB in free air at 55 Hz or about 40 Hz in-car. Adding a little boost will easily take it all the way out to 30 Hz in-car.
- The Kicker C10c-4 aligns acceptably in a 1-ft³ vented box, even though it's slightly undersized. Without boost, this system is down 3 dB at 40 Hz in free air and should hit 30 Hz in-car. With a slight boost, the system should be flat out to nearly 20 Hz in-car. Power handling with this setup will be excellent.
- The Kicker C12c-4 is misaligned in the 2-ft³ vented box. In free air, the response has a much bigger hump than is normally considered acceptable. However, our experience has shown that in-car response should be flatter. Unboosted, this system is down 3 dB at 35 Hz in free air, or high 20s in-car. With boost it should reach the low 20s. Again, expect terrific power handling.
- The MTX T584 aligns well in the 0.5-ft³ vented box. Expect the unboosted −3-dB point to be about 48 Hz in free air, which

should be close to 35 in-car. With boost, the system goes out to 27 Hz in free air. Just a trace of boost should take this system to 25 Hz in-car.

- The MTX T5104 aligns fairly well in a 1.5-ft³ vented box, although it is somewhat undersized. Still, the response hump is less than 1.0 dB, which is not that shabby. The unboosted −3-dB point is a very good 32 Hz in free air, which means it will reach into the mid 20s in-car and handle plenty of power.
- The MTX T7124 is misaligned in a 2-ft³ vented box. It really needs 3 or more cubic feet. The unboosted response has a 2-dB peak, but it is down 3 dB all the way out to 30 in free air. The filter smoothes the response quite a bit and extends out to 24 in free air. It should hit 20 Hz in-car. This system will easily handle hundreds of watts of power.

Figures 6-6 and 6-7 are the dual driver versions of Figures 6-4 and 6-5. Some of them are the hatchback designs of Chapter 3, repeated here for comparison purposes. As indicated in the tables, response curves for all of the systems in Figures 6-4 through 6-7 are shown in Figures 6-8 through 6-13.

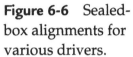

Figure 6-6 Sealed-box alignments for various drivers.

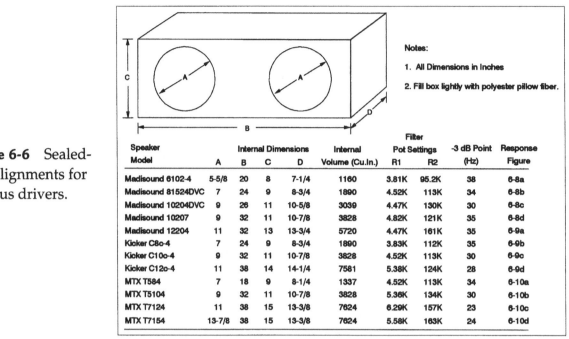

Notes:

1. All Dimensions in Inches
2. Fill box lightly with polyester pillow fiber.

Speaker Model	Internal Dimensions				Internal Volume (Cu.In.)	Filter Pot Settings		-3 dB Point (Hz)	Response Figure
	A	B	C	D		R1	R2		
Madisound 6102-4	5-5/8	20	8	7-1/4	1160	3.81K	95.2K	38	6-8a
Madisound 81524DVC	7	24	9	8-3/4	1890	4.52K	113K	34	6-8b
Madisound 10204DVC	9	26	11	10-5/8	3039	4.47K	130K	30	6-8c
Madisound 10207	9	32	11	10-7/8	3828	4.82K	121K	35	6-8d
Madisound 12204	11	32	13	13-3/4	5720	4.47K	161K	35	6-9a
Kicker C8c-4	7	24	9	8-3/4	1890	3.83K	112K	35	6-9b
Kicker C10c-4	9	32	11	10-7/8	3828	4.52K	113K	30	6-9c
Kicker C12c-4	11	38	14	14-1/4	7581	5.38K	124K	28	6-9d
MTX T584	7	18	9	8-1/4	1337	4.52K	113K	34	6-10a
MTX T5104	9	32	11	10-7/8	3828	5.36K	134K	30	6-10b
MTX T7124	11	38	15	13-3/8	7624	6.29K	157K	23	6-10c
MTX T7154	13-7/8	38	15	13-3/8	7624	5.58K	163K	24	6-10d

Figure 6-7 Vented-box alignments for various drivers.

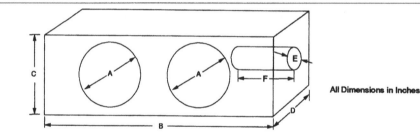

All Dimensions in Inches

Speaker Model	Internal Dimensions						Internal Volume (C.I.)	Box Freq. Fb (Hz)	Filter Pot Settings		-3dB Point (Hz)	Figure
	A	B	C	D	E	F			R1	R2		
Madisound 6102-4	5-5/8	20	8	7-1/4	3	12	1160	42	3.01K	109K	38	6-11a
Madisound 8154	7	32	8	7-1/2	3	13	1920	30	4.47K	130K	29	6-11b
Madisound 10207	9	32	11	10-7/8	4	15	3828	28	4.62K	155K	27	6-11c
Madisound 12204	11	36	14	14-3/8	4	8	7245	25	4.82K	174K	30	6-11d
Kicker C8c-4	7	24	9	8-3/4	3	12	1890	27	8.04K	104K	27	6-12a
Kicker C10c-4	9	32	11	10-7/8	4	12	3828	30	5.36K	134K	27	6-12b
Kicker C12c-4	11	36	14	14-3/8	4	8	7245	25	5.36K	156K	24	6-12c
MTX T584	7	32	8	7-1/2	3	13	1920	31	8.03K	104K	27	6-13a
MTX T5104	9	34	13	12-7/8	4	11-3/8	5690	25	8.37K	109K	24	6-13b
MTX T7124	11	36	14	14-3/8	4	8	7245	25	6.58K	164K	24	6-13c

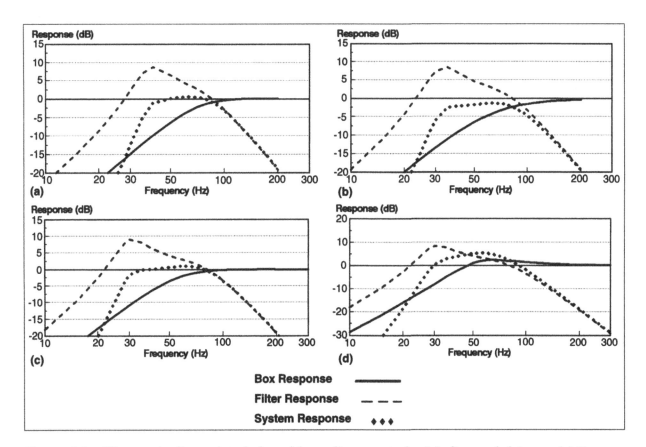

Figure 6-8 Electronically assisted closed-box alignments for Madisound drivers. (*a*) Two Madisound 6102-4s in 0.6-ft³ closed box. Q_{tc} = 0.78; Q_f = 2.5 at 35 Hz; R1 = 3.8 K; R2 = 95.2 K; F3 = 38 Hz. (*b*) Two Madisound 81524DVCs in 1.0-ft³ closed box. Q_{tc} = 0.6; Q_f = 2.5 at 32 Hz; R1 = 4.5 K; R2 = 113 K; F3 = 34 Hz. (*c*) Two Madisound 10204DVCs in 1.6-ft³ closed box. Both voice coils driven. Q_{tc} = 0.75; Q_f = 2.7 at 30 Hz; R1 = 4.47 K; R2 = 130 K; F3 = 30 Hz. (*d*) Madisound 10207 in 1.0-ft³ closed box. One voice coil driven. Q_{tc} = 1.2; Q_f = 2.5 at 30 Hz; R1 = 4.82 K; R2 = 121 K; F3 = 35 Hz.

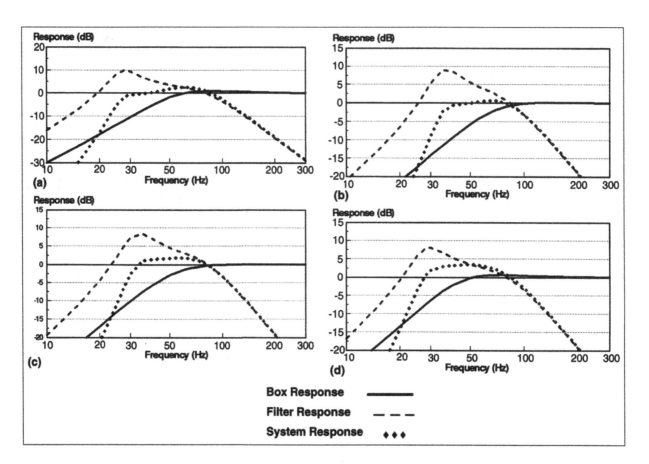

Figure 6-9 Electronically assisted closed-box alignments for various drivers. (*a*) Madisound 12204DVC in 1.5-ft³ closed box. Both voice coils driven. Q_{tc} = 0.94; Q_f = 3.0 at 27 Hz; R1 = 4.47 K; R2 = 161 K; F3 = 35 Hz. (*b*) Kicker C8c-4 in 0.5-ft³ closed box. Q_{tc} = 0.8; Q_f = 2.7 at 35 Hz; R1 = 3.83 K; R2 = 112 K; F3 = 35 Hz. (*c*) Kicker C10c-4 in 1.0-ft³ closed box. Q_{tc} = 0.76; Q_f = 2.5 at 32 Hz; R1 = 4.52 K; R2 = 113 K; F3 = 30 Hz. (*d*) Kicker C12c-4 in 2.0-ft³ closed box. Q_{tc} = 0.9; Q_f = 2.4 at 28 Hz; R1 = 5.38 K; R2 = 124 K; F3 = 28 Hz.

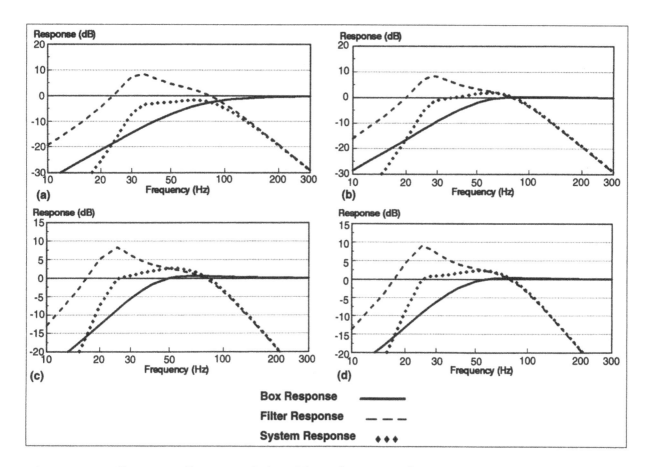

Figure 6-10   Electronically assisted closed-box alignments for MTX drivers. (*a*) MTX T584 in 0.35-ft³ closed box. $Q_{tc} = 0.62$; $Q_f = 2.5$ at 32 Hz; R1 = 4.52 K; R2 = 113 K; F3 = 34 Hz. (*b*) MTX T5104 in 1.0-ft³ closed box. $Q_{tc} = 0.84$; $Q_f = 2.5$ at 27 Hz; R1 = 5.36 K; R2 = 134 K; F3 = 30 Hz. (*c*) MTX T7124 in 1.3-ft³ closed box. $Q_{tc} = 0.89$; $Q_f = 2.5$ at 23 Hz; R1 = 6.29 K; R2 = 157 K; F3 = 23 Hz. (*d*) MTX T7154 in 2.0-ft³ closed box. $Q_{tc} = 0.83$; $Q_f = 2.7$ at 24 Hz; R1 = 5.58 K; R2 = 163 K; F3 = 24 Hz.

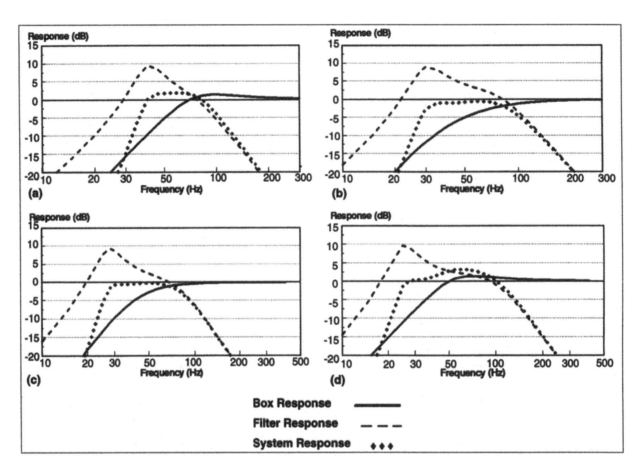

Figure 6-11 Thiele-Small sixth-order alignments for Madisound drivers. (*a*) Two Madisound 6102-4s in 0.6-ft³ vented box. $Q_f = 3.0$ at 40 Hz; R1 = 3.0 K; R2 = 109 K; $F_b = 42$ Hz; F3 = 38 Hz. (*b*) Two Madisound 8154s in 1.0-ft³ vented box. $Q_f = 2.7$ at 30 Hz; R1 = 4.47 K; R2 = 130 K; $F_b = 30$ Hz; F3 = 29 Hz. (*c*) Madisound 10207 in 1.0-ft³ vented box. $F_b = 28$ Hz; $Q_f = 2.9$ at 27 Hz; R1 = 4.62 K; R2 = 155 K; F3 = 27 Hz. (*d*) Madisound 12204DVC in 2-ft³ vented box. $F_b = 25$ Hz; $Q_f = 3.0$ at 25 Hz; R1 = 4.82 K; R2 = 174 K; F3 = 30 Hz.

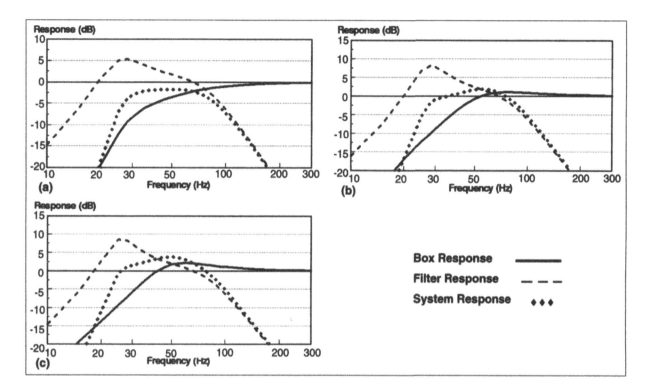

Figure 6-12 Thiele-Small sixth-order alignments for selected MTX drivers. (*a*) Kicker C8C-4 in 0.5-ft³ vented box. Q_f = 1.8 at 25 Hz; R1 = 8.0 K; R2 = 104 K; F_b = 32 Hz; F3 = 27 Hz. (*b*) Kicker C10C-4 in 1.0-ft³ vented box. Q_f = 2.5 at 25 Hz; R1 = 5.36 K; R2 = 134 K; F_b = 30 Hz; F3 = 27 Hz. (*c*) Kicker C12C-4 in 2.0-ft³ vented box. Q_f = 2.7 at 25 Hz; R1 = 5.36 K; R2 = 156 K; F_b = 25 Hz; F3 = 24 Hz.

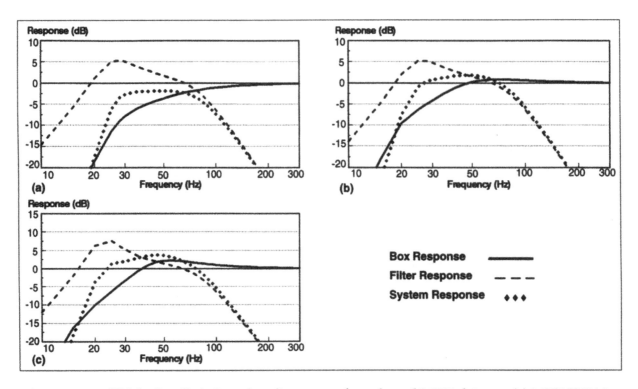

Figure 6-13   Thiele-Small sixth-order alignments for selected MTX drivers. (*a*) MTX T584 in 0.5-ft^3 vented box. Q_f = 1.8 at 25 Hz; R1 = 8.0 K; R2 = 104 K; F_b = 31 Hz; F3 = 27 Hz. (*b*) MTX T5104 in 1.5-ft^3 vented box. Q_f = 1.8 at 24 Hz; R1 = 8.37 K; R2 = 109 K; F_b = 25 Hz; F3 = 24 Hz. (*c*) MTX T7124 in 2.0-ft^3 vented box. Q_f = 2.5 at 22 Hz; R1 = 6.58 K; R2 = 164 K; F_b = 25 Hz; F3 = 24 Hz.

In Conclusion

In this chapter, we have attempted to show that there are many different alignments that we could envision for just the few drivers chosen as our examples. Space, power handling, and cost are the primary parameters that seem to drive the decision-making process. Hopefully, these examples will help you to make the right selection for your application.

Chapter 7

The Filter Crossover

At the heart of each of the biamplified systems in this book is the filter crossover. This circuit controls the subwoofer and makes it possible for the main speaker system to operate without bass boost. The filter performs five functions, which are listed as follows:

1. *Summing right and left channels.* The first step in the process is to extract the bass information from both the right and left channels and mix it together in such a way as to not cause electrical interaction between the power amplifiers in the dash unit. While only two summing inputs are required to do this, this filter is provided with four (a somewhat unique feature in itself). By connecting them to all four front-rear, right-left head unit outputs, bass level is rendered insensitive to right-left balance *and* front-rear fader positions.

   With this filter, you can select either high- or low-level input sources. On some aftermarket head units, connection to the filter can be done more conveniently by utilizing the low-level outputs. However, it should be noted that you will encounter fewer problems with ground loops by using the high-level outputs.

   While the filter is compatible with virtually all auto sound equipment (and home equipment as well), **the integrated** circuit will be damaged by connecting it directly to the speaker connections on a high-powered amplifier. When necessary, the filter can still be used in that application by installing a "high-to-low level adapter" which is available at any auto sound dealer. The adapter will be required in virtually all high-level home applications.

2. *Gain control.* One of the two external controls on the filter. The gain control provides adjustment of the subwoofer volume relative to the main speakers.

3. *Low-frequency tuning.* The circuit board has two adjustable trimpots. Together they set the *Q* or damping factor and the low-frequency cutoff point. Their settings are customized for each particular system. A number of example alignments have been given for comparison, including some for drivers, which were designed specifically for auto sound competition.

4. *Rumble filter.* The circuit also provides for built-in rumble suppression, which enhances low-frequency definition or clarity. This function also helps to create what is referred to as "tight bass."

5. *Electronic crossover.* The electronic crossover attenuates the higher frequencies so that only low bass reaches the subwoofer. The second external control allows you to set the crossover frequency point between approximately 30 and 330 Hz in order to get the best match between your main drivers and the subwoofer. The crossover slopes are designed to provide optimum balance between maximum bass attack and attenuation of higher-frequency information. Properly tuned, the system will be both "fast" and "tight."

Clarity

As stated in the introduction, removing low-frequency workload from the main drivers in an auto sound system improves the system clarity and power handling. This is because reproducing low frequencies requires the movement of large quantities of air. If the driver cone diameter is small, the cone excursions necessary to move enough air become impossibly large. For this reason, the minimum diameter of what can accurately be called a woofer is about 6 in and even then a 6-in woofer that is attempting to reproduce 40-Hz tones at any reasonable volume will probably not be able to reproduce midrange (500 Hz to 1 kHz) clearly.

To make matters worse, most factory systems utilize full-range drivers, which means that a 4-in dash-mounted driver is attempting to reproduce 40- and 15,000-Hz tones at the same time. It is usually obvious to even the casual listener that turning up the bass in a conventional auto sound system causes gross amounts of distortion.

So enter the subwoofer. It has one purpose—to independently reproduce the low-bass information—which amounts to about 5 percent of the audio spectrum—in order to free up the smaller main drivers to do their job of dealing with the other 95 percent without the added stress of boosted bass. In so doing, the workload on both the main drivers and head unit is reduced and everything sounds clearer.

Now, some of you may be confused at this point or even disagree that simply not boosting the bass is a legitimate way of cleaning up your system. However, there are two things working together to make it happen. First, the main drivers are, in fact, designed to handle a 12-W-per-channel, full-range unboosted signal by mechanically absorbing the low-bass signal content. In other words, they will simply ignore it as long as it's not boosted. In the process, they will cleanly respond to the parts within their capability, which will generate the very important leading edge of the bass transients. A second little-known fact is that it takes 5 to 6 times more power to reproduce low bass than to reproduce the middle frequencies. Therefore, a 12-W-per-channel head unit already makes a pretty good match for a 100-W bass amplifier.

All we need to do then is to extract the low-bass signals from both right and left channels and direct them to a separate amplifier and speaker system designed specifically for this purpose. The rest should be straightforward. This can be done with a simple "low-pass filter" which is built into most recent amplifiers as an added feature. But there's much more to our story than this. The filter we'll be building is considerably different from the ones you can buy commercially. Remember the five functions it performs? Number 3 is the critical one that sets it apart and is just not available anywhere else. It is the feature that allows us to obtain clean, low bass from small enclosures. Taken altogether, the controls built into the filter allow for adjustments to compensate for differences in passenger compartment volume and efficiencies of the various main drivers, and match whatever their natural low-frequency roll-off point happens to be. We can even "dial in" the filter to compensate for manufacturing variations encountered in the bass drivers themselves. Last but not least, the overall system sound can be adjusted to suit individual taste. Many users have commented about the fact that if a recording didn't have enough bass, "you can make your own."

The filter schematic diagram is shown as Figure 7-1, and an enlarged drawing of the actual circuit board layout is shown as Figure 7-2. The circuit employs a high-performance quad op amp chip to perform all of the functions and can be one of several different types. I chose the LF-347, although the TL084 is an acceptable substitute. The circuit is fairly easy to build in spite of its apparent complexity, and a number of people across the country have successfully built it. For those of you who find the task too formidable or just are not inclined, a fully assembled and tested unit made from high-quality components is available from the author (dferg1000@aol.com) for approximately $100.

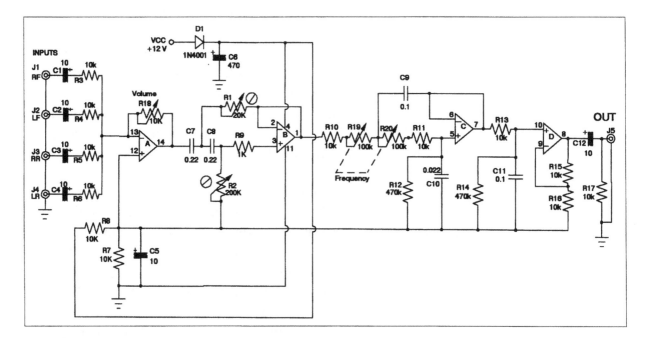

Figure 7-1 Subwoofer filter schematic.

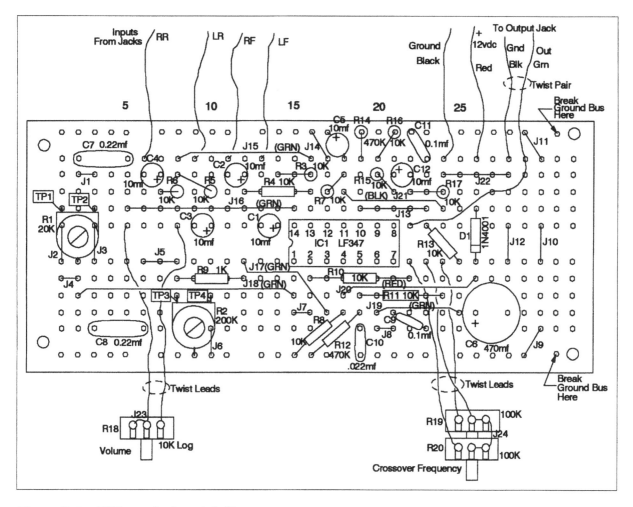

Figure 7-2 PC board pictorial diagram.

Circuit Function

Referring back to Figure 7-1, op amp A sums all the inputs with pot R18 setting the gain of this section from 0 to 1.0. You can use either two or four inputs, depending on your particular hookup, and the order of the input connections has no importance.

Op amp B, in conjunction with the *RC* network C7, C8 and trim-pots R1 and R2, form the tunable second-order high-pass filter. The actual trimpot values selected determine both the Q, which is the inverse of this section's damping factor, and the corner frequency F_c, which is approximately equal to the boost peak frequency F_b. This can all be calculated with the following equations (Ref. 1):

$$K = 10^{B/20} \tag{7-1}$$

where B equals the desired boost in dB

$$Q = \{[K^2 + K(K^2 - 1)]/2\}^{1/2} \tag{7-2}$$

$$F_c = F_b[1 - 1/(2Q^2)]^{1/2} \tag{7-3}$$

For this particular circuit, the following equations apply:

$$R_1 = 1/(4\pi F_c QC) \tag{7-4}$$

where $C = C_7 = C_8$

$$R_2 = 4Q^2 R_1 \tag{7-5}$$

For convenience, Table 7-1 summarizes the range of possible settings, while the equations above will allow you to calculate exact values in between those listed. This is illustrated graphically as the area between the two curves in Figure 7-3.

The subwoofer crossover function is performed by sections C and D and is also somewhat unique in that it has variable slopes. Section C is a second-order low-pass filter with a continuously variable corner frequency range of 30 to 330 Hz set by the position of dual pots R19 and R20. It is followed by section D, which is a first-order low-pass filter with a fixed corner frequency of 159 Hz. Taken together, sections C and D form a third-order low-pass filter. The values selected are intended to provide maximum bass attack while imparting apparent "smoothness." In more practical terms, you can crank it up and punch up the bass line of most recordings without its sounding like rolling thunder coming down the road. It will sound more, ahem . . . natural. When properly adjusted, the subs will blend right in and give the illusion that the midbass is coming from the front main drivers, which are usually a little weak in that area.

Table 7-1 Subwoofer Trim Pot Settings for Various Boosts and Frequencies

Freq (Hz)		22		24		26		28		30		32		34		36		38		40	
	Boost	Trim Pot Settings (K Ohms)																			
Qf	(dB)	R1	R2	R1	R2	R1	R2	R1	R2	R1	R2	R1	R2	R1	R2	R1	R2	R1	R2	R1	R2
0.71	0					19.7	39.3	18.3	36.5	17.1	34.1	16.0	32.0	15.0	30.1	14.2	28.4	13.5	26.9	12.8	25.6
1.13	2	14.6	74.2	13.4	68.0	12.3	62.8	11.4	58.3	10.7	54.4	10.0	51.0	9.4	48.0	8.9	45.4	8.4	43.0	8.0	40.8
1.30	3	12.6	85.8	11.6	78.7	10.7	72.6	9.9	67.4	9.2	62.9	8.7	59.0	8.2	55.5	7.7	52.4	7.3	49.7	6.9	47.2
1.49	4	11.0	98.2	10.1	90.0	9.3	83.1	8.7	77.2	8.1	72.0	7.6	67.5	7.1	63.6	6.7	60.0	6.4	56.9	6.1	54.0
1.70	5	9.7	111.8	8.9	102.5	8.2	94.6	7.6	87.8	7.1	82.0	6.7	76.8	6.3	72.3	5.9	68.3	5.6	64.7	5.3	61.5
1.93	6	8.5	126.7	7.8	116.2	7.2	107.2	6.7	99.6	6.3	92.9	5.9	87.1	5.5	82.0	5.2	77.4	4.9	73.4	4.7	69.7
2.18	7	7.5	143.3	6.9	131.4	6.4	121.3	5.9	112.6	5.5	105.1	5.2	98.5	4.9	92.7	4.6	87.6	4.4	83.0	4.2	78.8
2.46	8	6.7	161.7	6.1	148.3	5.7	136.9	5.3	127.1	4.9	118.6	4.6	111.2	4.3	104.7	4.1	98.8	3.9	93.6	3.7	89.0
2.77	9	5.9	182.3	5.4	167.1	5.0	154.3	4.7	143.2	4.3	133.7	4.1	125.3	3.8	118.0	3.6	111.4	3.4	105.6	3.3	100.3
3.12	10	5.3	205.3	4.8	188.2	4.5	173.7	4.1	161.3	3.9	150.5	3.6	141.1	3.4	132.8	3.2	125.5	3.0	118.9	2.9	112.9
3.51	11					4.0	195.4	3.7	181.5	3.4	169.4	3.2	158.8	3.0	149.5	2.9	141.1	2.7	133.7	2.6	127.0
3.95	12							3.3	204.1	3.1	190.5	2.9	178.6	2.7	168.1	2.5	158.7	2.4	150.4	2.3	142.8

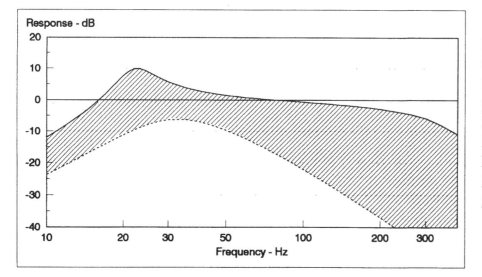

Figure 7-3 Range of subwoofer filter tuning. When high pass $Q = 3$, $F_c = 22$ Hz, low pass $F_c = 330$ Hz. When high pass $Q = 0.71$, $F_c = 40$, low pass $F_c = 30$ Hz.

In the following parts list, you will note that the parts to make one filter come from three different suppliers. Unfortunately, it is difficult to get exactly the right parts from a single source. Therefore, I had to resort to getting the best parts from the best suppliers. As an example, you will note that the control pots come from Radio Shack. This is because, for many years, they have sold a small, high-quality, dual 100K pot that is difficult to get anywhere else.

Parts List

Part	Value	Quantity	Part No.	Supplier
Resistors:				
R3 through R8, R10, R11, R13 through R17	10K	12	10.0KX	Digi-Key
R9	1K	1	1.00KX	Digi-Key
R12, R14	475K	2	475KX	Digi-Key
R1	20K trimpot	1	K4A24	Digi-Key
R2	200K trimpot	1	K4A25	Digi-Key
R18	10K pot	1	271-1721	Radio Shack
R19, R20	100K dual pot	1	271-1732	Radio Shack
Diodes:				
D1	1N4001	1	1N4001GI	Digi-Key
Capacitors:				
C1 through C5, C12	10 mf/35 V electrolytic	6	P5253	Digi-Key
C6	470 mf/35 V electrolytic	1	P5260	Digi-Key
C7, C8	0.22 mf/50 V metal film	2	P4667	Digi-Key
C9, C11	0.1 mf/50 V metal film	2	P4525	Digi-Key
C10	0.022 mf/50 V metal film	1	P4517	Digi-Key
Op amps:				
A,B,C,D	LF-347	1	LF347N	Digi-Key
Hardware:				
Circuit board		1	276-170	Radio Shack
Knobs	¾-in diameter	2	274-415	Radio Shack
Cabinet	5½ by 3 by 1¼ in	1	537-139-P	Mouser
RCA jacks		5	161-1052	Mouser
Grommet	⁵⁄₁₆-in diameter	1	5176-211	Mouser
Nylon spacers	¼ in long	4	561-K4.25	Mouser

Parts List (*Continued*)

Part	Value	Quantity	Part No.	Supplier
Screws	4-40 by ½ in	4	H146	Digi-Key
Nuts	4-40	4	H216	Digi-Key
IC socket	14 pin	1	A9314	Digi-Key
Wire:				
Red, green, black	22 gage stranded	1 package	278-1224	Radio Shack
Bus wire	24 gage	1 spool	278-1341	Radio Shack

Total cost for all parts is approximately $35 including shipping. See the Resource Directory at the back of the book for addresses and phone numbers. You may want to order a few extra of the inexpensive parts just in case something goes wrong. Many of these parts are pretty small and could get lost or damaged in the assembly process.

Tools

You will need the following tools:

1. Electric drill and ⅛-, ¼-, and ⁵⁄₁₆-in bits
2. 30-W pencil-type soldering iron
3. Rosin core 60/40 solder, approximately 0.032-in diameter
4. Solder wick (desoldering braid)
5. Small diagonal wire cutters
6. Wire strippers
7. Hacksaw
8. Small screwdriver
9. Needlenose pliers
10. File
11. Volt-ohmmeter (high-impedance digital type preferred)

Chassis Preparation

1. In order to fit the circuit board into the cabinet, you must first cut off approximately 2 in from its right end. (See Figure 7-2.) Using a fine-toothed hacksaw, carefully cut across the board at line 35. To avoid splintering the board, support it on a flat surface and position the hacksaw blade *parallel* to the board surface while cutting. File all edges smooth.

2. Enlarge hole number 32 on both top and bottom busses with a ⅛-in bit. These holes will be used with the two holes provided on the left end of the board to bolt the circuit board to the cabinet bottom.

3. Using a ¼-in drill bit, partially drill out hole number 31 on both busses so as to just break the foil. This will prevent the busses from shorting to ground when the circuit board is bolted to the cabinet.

4. Separate the two halves of the cabinet by prying with a pocket knife, being careful not to lose the two small screws inside. The piece with the end flaps is the top while the piece with sides will be the bottom.

5. Place the circuit board in the bottom of the chassis as shown in Figure 7-4a and position it to within about ¹⁄₁₆ in from the rear flap. Mark the four chassis holes carefully with a sharp pencil.

6. Center-punch the four marks and drill the four mounting holes in the bottom of the chassis with a ⅛-in bit.

7. Place the board back in the cabinet bottom and verify that all the holes are aligned by test fitting all four mounting screws.

8. Remove the circuit board and set aside. Referring to Figure 7-4b, locate and center-punch the mounting holes for the control potentiometers. Pilot drill all holes with a ⅛-in bit and then enlarge the two control shaft holes to ⁵⁄₁₆ in. Ream out the holes with successively larger bits to avoid problems with the drill seizing. Test-fit both controls and then set aside.

9. Referring to Figure 7-4c, drill the six holes in the rear of the cabinet using the same procedure as in step 8. Be sure to remove all burrs.

10. Using a hacksaw, cut off both potentiometer shafts to ⁵⁄₁₆ in beyond the threaded mounting sleeve. File the ends smooth and test-fit the control knobs. If you decide to use control knobs larger than those specified, or if you are planning to mount the filter on the rear of a panel, you may wish to leave more shaft extension. The prepared hardware is shown in Photo 7-1.

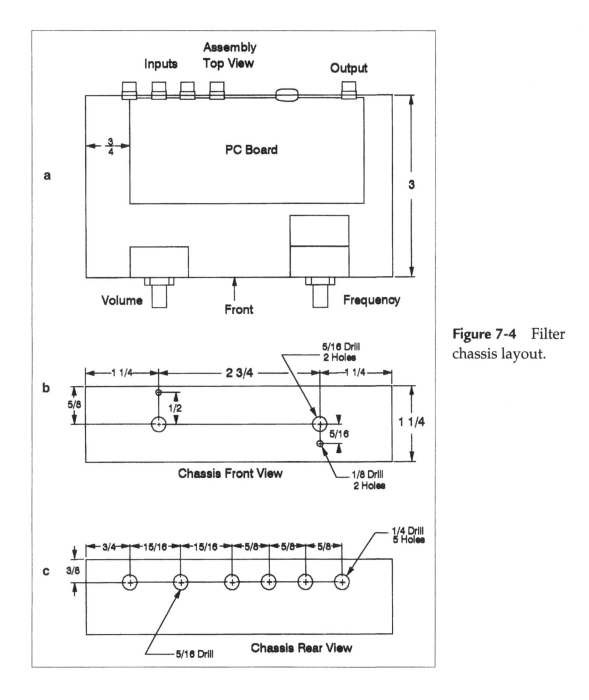

Figure 7-4 Filter chassis layout.

Photo 7-1

11. Wire the potentiometers as shown in Figure 7-2. The view shown is looking down at the pots with their solder lugs facing up. Begin by cutting appropriate lengths of bus wire; then install jumpers J23 and J24. Secure the jumpers by soldering them at each terminal not shared with a lead wire. Then cut five pieces of assorted color stranded wire to 6-in lengths, strip one end, and wire and solder to the potentiometers. After soldering to the pots, pull wires together and trim to make all three lengths about 3½ in. Strip the new ends evenly and twist carefully. The completed pots:

- Test the 10K volume potentiometer wiring by hooking up an ohmmeter across its leads. Rotate the pot fully counterclockwise and verify that the resistance drops to zero. Rotating it fully counterclockwise should cause the ohmmeter reading to increase to approximately 10K ohms.

- Test the dual 100K pot by placing the ohmmeter across the front pot and common center lead. Rotate the pot counterclockwise, and the ohmmeter reading should increase to 100K. Leave one ohmmeter lead on the center pot lead and transfer the other ohmmeter lead to the rear pot lead. The resistance should read approximately the same as for the front pot. Next, rotate the pot clockwise until the ohmmeter reads approximately 10K. Move the ohmmeter lead to the front pot and verify that it also reads approximately 10K. If it does, you're ready to proceed with circuit board assembly.

Assembly

Referring back to the pictorial diagram of Figure 7-2, assemble the electronic components on the board. Use the exact hole locations shown. Solder each point carefully. The following suggestions should help.

1. Avoid excessive heat to prevent damage to the circuit board foil. Keep the soldering iron in contact with the wire leads and foil the minimum time required to completely flow the solder. Touch the solder to the foil or component lead, not to the iron's tip. For best results, use a small conical soldering iron tip. If you use an inexpensive one like the Radio Shack 64-2067, it's a good idea to sharpen the tip with a file before proceeding with assembly.
2. Clean the iron frequently with a damp sponge to remove built-up contaminants.
3. If you make a mistake, unsolder it with desoldering braid. However, be careful, as it is easy to apply excessive heat during desoldering.
4. After soldering, clip leads flush with the solder beads using diagonal wire cutters.
5. When the board is completely finished, remove residue with flux remover (like Radio Shack Catalog No. 64-4330). Do this outdoors, as the solvents are volatile and will damage painted surfaces.

Assembly Order

Start by recognizing the location of the IC socket as "home base" and familiarize yourself with the relative positions between it and the other components. The following order of assembly and in-progress photos should help in minimizing errors:

1. Install all of the bare bus wire jumpers J1, J2, J3, J4, J5, J6, J7, J8, J9, J10, J11, J12, J13, J14, and J22 (Photo 7-2).

Photo 7-2

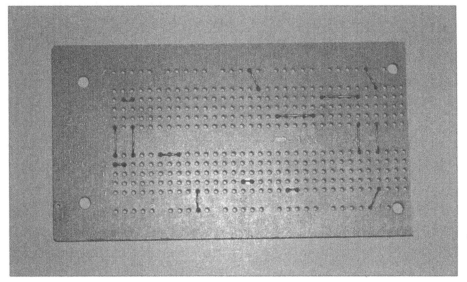

2. The 14-pin IC socket
3. Resistors R3, R4, R5, R6, R7
4. Capacitors C1, C2, C3, C4
5. Green jumpers J15, J16
6. Trimpots R1, R2
7. Capacitors C7, C8 (Photo 7-3)

Photo 7-3

8. Resistor R9
9. Green jumpers J17, J18
10. Resistors R8, R10, R11, R12
11. Capacitors C9, C10
12. Green jumper J19, red jumper J20
13. Resistor R13, diode D1
14. Resistors R14, R15, R16, R17
15. Capacitors C12, C11
16. Black jumper J21
17. Capacitor C6 (Photo 7-4)

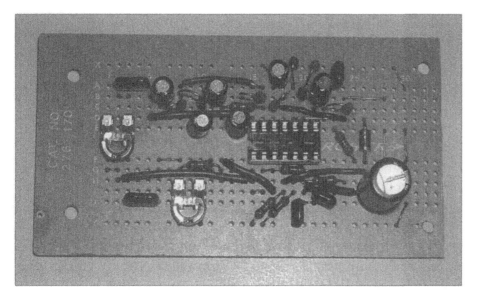

Photo 7-4

18. All four input leads
19. Red and black power leads (approximately 18 in long)
20. Green and black output leads
21. Volume control leads from R18
22. Crossover frequency control leads from R19 and R20 (Photo 7-5)

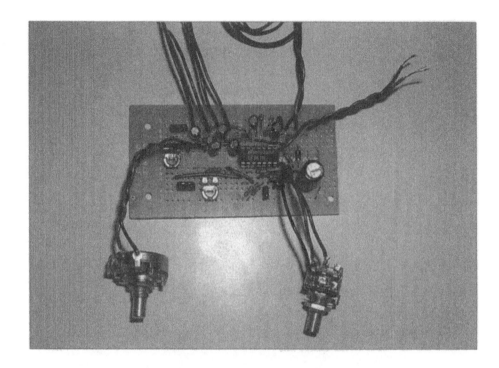

Photo 7-5

Before inserting the IC chip, test the integrity of the circuit by performing a dc voltage check. Connect a convenient dc voltage to the power leads. For the sake of discussion, we will assume it is 12 V. With your volt-ohmmeter set to measure dc volts, connect the black meter lead to either of the black ground wires and touch the red lead to any of the jumpers J9, 10, or 11. Turn on the supply voltage and you should be able to watch the voltage at jumper J9 slowly climb to one-half the supply voltage less 0.7 V drop across the diode. For a 12.0-V dc supply the bus voltage should be 5.65 V.

Using the test probe tips, touch jumper J9 and J7 at the same time. The meter should indicate 0.0 V. Do the same for jumpers J9 and J8. The results should again be 0.0 V.

Next, turn off all power and discharge the capacitors by touching jumper J9 with one of the black power leads for several seconds. You are now ready to insert the IC in the socket except for one precaution. The LF-347 chip can cost $2 or more, depending upon where you bought it. Also, it may not be readily available locally without paying $5 or more. If you are a new experimenter, you may want to do your initial circuit testing with some inexpensive LM-324s, which are available at Radio Shack. While these could be considered sacrificial, they will work in a pinch even though they don't have nearly the rated performance of the LF-347s.

When you are inserting the chip, pay close attention to be sure all the pins are properly aligned. You may have to squeeze them together a little to get everything to fit. Also be absolutely certain that the notch in the end of the chip is oriented correctly (to the left). Otherwise, when you turn on the power you will let all of the smoke out of the chip, which can cause considerable anguish if you waited a week to get one in the mail.

OK . . . you made it this far. The chip's in, the power's on, and no smoke. Take your voltmeter again and carefully (touching only one pin with each probe) measure across pins 12 and 14, 1 and 14, 1 and 7, and 7 and 8. In each case, the reading should be 0 V. If all this checks out, you're ready to hook the filter up for testing. Incidentally, if any of these voltages were out of balance, don't waste time going any further without resolving the problem.

Bench Test

If you have test equipment, now is the time to check your circuit out with a signal generator and oscilloscope. Connect the signal generator to one of the input wires and the black power ground wire. Connect the filter output green and black pair to the scope. Apply a convenient dc voltage (between 12 and 30 V) to the red and black pair of power wires. For test purposes, set trimpots R1 and R2 at 6K and 96K, respectively. Do this by simply placing the tips of your ohmmeter test probes on the metal tabs on the trimpots at the test points indicated as TP1-TP4 in Figure 7-2. Rotate each adjustment screw with a small screwdriver until the desired resistance is obtained. Then set both external controls fully clockwise. Sweep through the frequency range of about 20 to 500 Hz with the signal generator. The output waveform should be clean and sinusoidal. Its amplitude should start low as seen in Photo 7-11a and increase with frequency to a peak at around 32 Hz as seen in Photo 7-11b, then fall to a brief plateau before falling rapidly to near zero at the higher frequencies as seen in Photo 7-11e. Referring back to Figure 7-3, the upper curve is representative of the general shape.

With the frequency set for a moderate filter output, transfer the input signal to each of the input leads in turn and verify that each input is functioning normally. Touch any of the remaining input leads to the one with the signal generator, and the output amplitude should double. Touch a third and the output should triple. Touch a fourth and . . . you get the idea. Then rotate the volume control counterclockwise. The output should fall to zero. Return the volume control to the fully clockwise position and then rotate the frequency control fully counterclockwise. The signal should be reduced only moderately and should remain clean.

Final Assembly

If everything checks out up to this point, you're ready for final filter assembly into the cabinet. For those without test equipment, you'll have to proceed somewhat on faith, so check and double-check all wire and solder joints one last time. A magnifying lens and strong light will also help to spot defects.

For a finished look, you can make front and rear "faceplates" using the full-sized graphics given in Figure 7-6. Photocopy or scan them onto full-page blank label stock (like Avery 8255). Make several copies in case you make a mistake. First cut the labels to rough size, leaving some border to trim. Carefully cover the face with wide cellophane tape; then trim to final size. Clean the cabinets with alcohol to remove any fingerprint residues and then carefully apply the labels to the front and back. Make the cutouts for mounting the components with a very sharp razor tool like a wallpaper knife. If the tool is the least bit dull, everything will tear.

Install all of the jacks and the rubber grommet. Position the grounding lugs on the jacks as shown in Figure 7-5; then "daisy chain" a length of bus wire through each of them. Solder all the lugs except the output jack. The completed cabinet is shown in Photos 7-6 and 7-7.

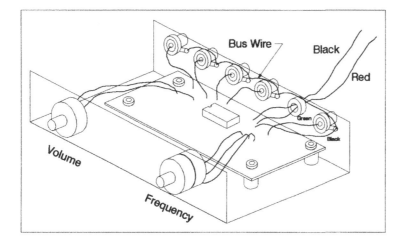

Figure 7-5 Filter chassis assembly drawing.

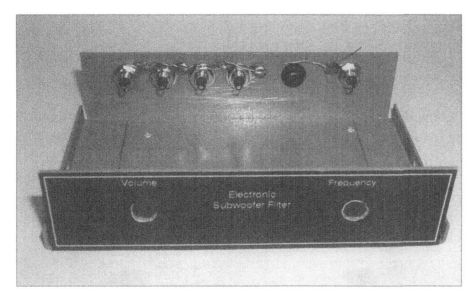

Photo 7-6

Photo 7-7

Now place the circuit board in the cabinet. Referring to Figure 7-5, twist the leads to both control pots and route them neatly to the front of the cabinet. Install the washers and nuts on the controls and tighten.

Trim the black and green output pair of wires to a minimum practical length and twist neatly together. Solder the green lead to the output jack center and the back to the last solder lug. This will completely ground the chassis to the rest of the circuit.

Pair up the black and red power leads and tie them in a knot near the grommet to act as a strain relief. Then pass the leads through the grommet.

Now trim each of the four input leads to length and solder to the center of each of the input jacks.

Finally, install the four spacers under the circuit board and bolt it in with four of the screws and nuts.

Photo 7-8 shows the completed filter without the top. The finished product is shown in Photos 7-9 and 7-10. The commercial version is very similar but more compact—an inch smaller in width and depth.

Photo 7-8

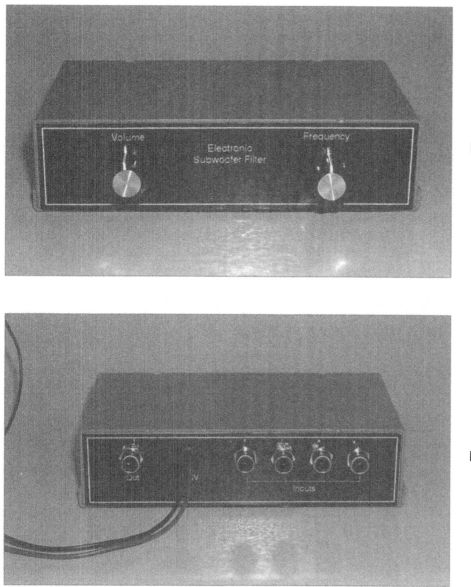

Photo 7-9

Photo 7-10

Performance

A properly assembled filter is both rugged and reliable and has an extremely clean output. To illustrate this, Photo 7-11 shows the output waveforms at selected frequencies. In all of the photos, the filter output is the upper trace and the signal generator is the lower trace. For this series, R1 and R2 were set at 6K and 96K, respectively, corresponding to a high-pass filter Q of 2.0 (6-dB boost) at 30 Hz. Also, the crossover frequency control was set at maximum, or fully clockwise.

In Photo 7-11*a*, at 15 Hz, the filter has a reduced output, illustrating the rumble-suppression feature. At 32 Hz, the filter is at its theoretical peak output for these settings, as shown in Photo 7-11*b*. It appears perfectly sinusoidal, as do all the others. At 100 Hz, the output is again clean but reduced, as shown in Photo 7-11*c*. At 200 Hz, output is significantly less in Photo 7-11*d*. At 1,000 Hz, the output is almost a flat line as shown in Photo 7-11*e*. To illustrate the low noise capability of the circuit, Photo 7-11*f* is the same conditions as Photo 7-11*e*, except the upper trace sensitivity has been set to the maximum on this particular oscilloscope—0.01 V/cm. Here the waveform is just as clean as all of the others at 100 times magnification!

Photo 7-11*a*
(15 Hz)

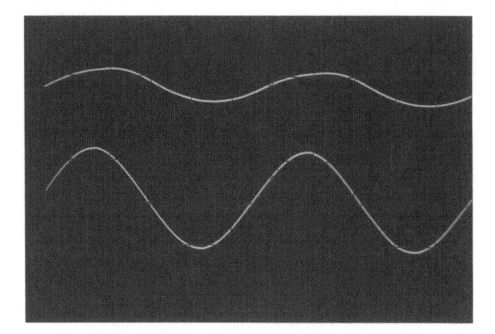

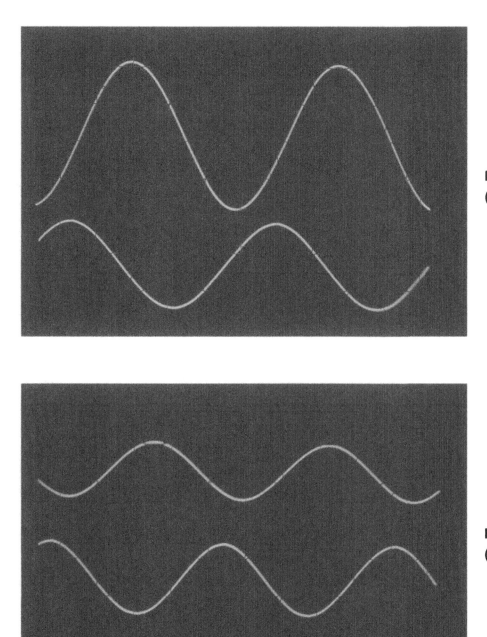

Photo 7-11*b*
(32 Hz)

Photo 7-11*c*
(100 Hz)

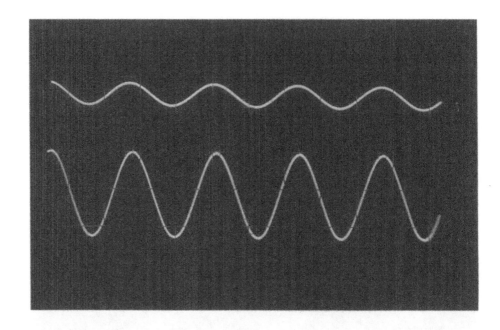

Photo 7-11d
(200 Hz)

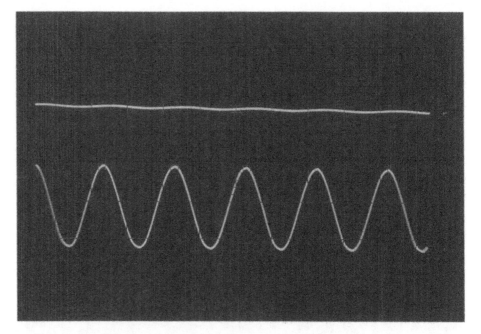

Photo 7-11e
(1000 Hz)

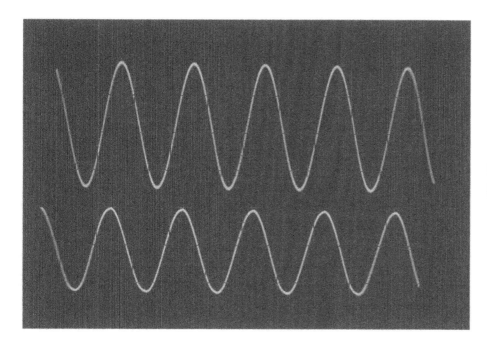

**Photo 7-11ƒ
(100 Hz)**

Testing in Use

Assuming you have no test equipment, the only other way to test the unit is in your car. To do this, connect the entire circuit shown in Figure 7-7. If you have no other convenient 12-V dc power supply, use

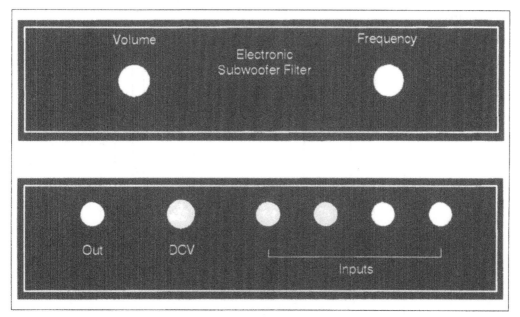

Figure 7-6   Actual size subwoofer filter enclosure graphics.

your car's battery to provide power. For connectors, use alligator clips or battery jumper cables.

Connect the crossover output with shielded cable to one or both inputs of your power amplifier, depending on which type you are using. Refer to Figure 7-8 to determine the appropriate hookup. Note that if you have selected a power amplifier which is not mono-

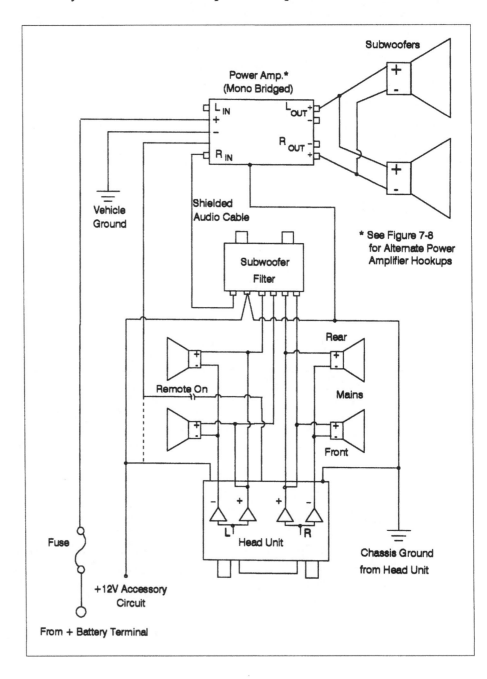

Figure 7-7 Auto wiring diagram.

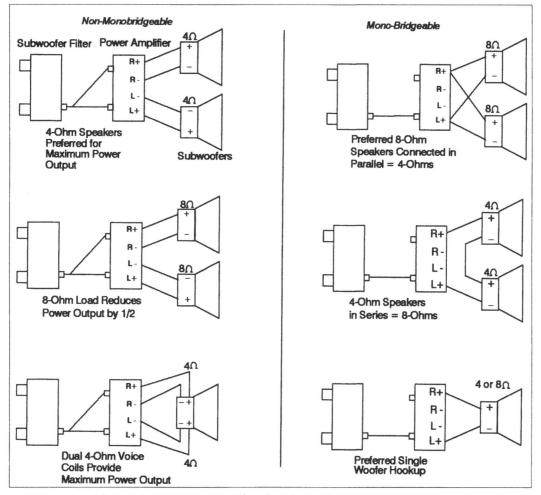

Figure 7-8 Alternate power amplifier hookups.

bridgeable, you will need a Y adapter like the Radio Shack No. 42-2435 to send the output to both amplifier inputs.

A convenient point to pick up a test signal is usually one of your car's rear speakers, especially if you have a sedan. In any case, connect a single wire between one of the dash unit's positive speaker outputs and one of the crossover inputs. For this hookup, you will have to make an adapter to go from a stripped wire to a phono plug. After checking all connections, power the system up and listen to the subwoofer. If it produces nothing but bass, congratulations! If you are testing with an unmounted driver, understand it cannot produce much bass without the proper baffling.

Test filter operation as follows. First turn the volume control clockwise and the volume should increase. Turn the frequency control clockwise and you should hear an increase in higher-frequency content, including the presence of vocals.

Finally, connect another single wire from the remaining positive rear driver output to one of the other crossover inputs. When you plug it in, the subwoofer volume should immediately increase. If it decreases, you have the wrong polarity speaker output lead. Go back and reconnect to the opposite lead and try again. This time, the volume should increase. Perform this same test on each of the remaining jacks. If you've made it successfully this far, the rest is easy.

Of course all of the preceding tests could have been performed in a low-level system application using shielded audio cables connected to the rear of the dash unit. In many cases, this is a much easier method as polarities are automatically correct. On the negative side, you may encounter more problems with ground loops using low-level hookups with this filter.

Caution

While you are doing all of your in-car testing, a word of caution is necessary. Never connect an auto stereo output lead to ground. Most modern auto dash units employ bridged output amplifiers, as was discussed in Chapter 2. This means that both positive and negative leads have the same signal present with respect to ground. The only difference between the two is that the negative lead is 180 degrees out of phase with the positive lead. If you touch either the positive or the negative lead of a bridged amplifier to ground for any length of time, it will destroy the output chip. In most cases that means the end of your deck, since they really don't make aftermarket decks to be repairable.

Finally . . .

At this point you're in the home stretch. All that remains is to get everything permanently installed in your car and clean up.

Chapter 8

Basic Speakerbuilding

Overview

In all of the construction projects presented in past chapters, the design work was done "behind the scenes." This is because, to the uninitiated, the science of speakerbuilding can be somewhat complex, and much of it is still considered "art." While I certainly don't claim to be an authority on the subject, I will attempt, in this chapter, to pass on some lessons learned. This will be in the form of an abbreviated road map and some hands-on pointers that should make it easier for those of you who are interested in learning more about loudspeaker design. However, in order to really get into this subject, you'll have to do quite a bit more reading, so along the way, sources of authoritative and practical reference literature will be provided. In my opinion, the single best resource for practical speakerbuilding is Ref. 3, *The Loudspeaker Design Cookbook* by Vance Dickason. The early issues of *Speakerbuilder* magazine are also a gold mine of information. Last but not least is David Weems's practical book (Ref. 11).

Part of the loudspeaker design process involves making some basic electrical and acoustic measurements. I will attempt to show you how to make them with minimum equipment, cost, and frustration. Once the measurements are done, there are calculations to do.

The easiest way to do them is with a personal computer and a loud-speaker design program. Internet addresses will be provided where you can download several useful shareware versions, and some sample printouts have been included to give you a feel for how these programs look. I will also provide you with the spreadsheet setup used to model my subwoofer filter response and its application in closed and vented box designs. Then, if you build (or purchase) the filter, you will be able to predict the results and make better decisions about selecting the right system for your particular need. Last but not least, we will apply all of the principles to a couple of designs and demonstrate how they work.

Thiele-Small Parameters

In order to design an enclosure or box for a woofer system, you must first have a working knowledge of the specifications that describe the physical and electrical properties of loudspeakers. Known today as Thiele-Small parameters (f_S, Q_{TS}, and V_{AS}), these were invented in the early 1970s by the now-famous audio engineer, A. N. Thiele, and subsequently refined by R. H. Small (Ref. 4), who took into account box losses by adding a new parameter Q_L—the box loss factor. The use of Thiele-Small parameters makes it possible to model a loudspeaker/box combination as a high-pass filter and get predictable results. Prior to this development, loudspeaker design was literally done by trial and error.

If you browse nearly any loudspeaker catalog, you will find the Thiele-Small parameters listed for each woofer. In those rare instances where they have been omitted, it is because, I'm told, the catalog publishers don't wish to confuse people with them. Instead of Thiele-Small parameters, they simply provide the enclosure specifications—a poor substitute in my opinion.

While knowing the published Thiele-Small parameters for a particular driver is a good starting point for your design, manufacturing variations can cause them to vary by as much as 25 percent. Therefore, if you build an enclosure without first measuring the woofer parameters, it could wind up being misaligned quite a bit and the resulting system will have an uneven bass response. One of the things we need to be able to do as speakerbuilders is to accurately measure Thiele-Small parameters. In order to do that, we have to be able to measure the driver's voice coil dc resistance and ac impedance for a given frequency. Think of impedance as resistance to alternating-current (ac) flow that varies with frequency. And unlike direct-current (dc) resistance, it has both a magnitude and phase angle; i.e., the current is usually not in phase with the voltage and the amount of phase lead or lag varies with frequency.

Impedance Measurement

Figure 8-1 shows the test setup referred to in the literature (Refs. 3, 5, and 11) as the "constant-current method" for measuring loudspeaker impedance. This particular setup has an added switch that makes the measurement process easier and more accurate and is usually not found in the literature. The switch must be of fairly high quality and have no measurable contact resistance in order for it not to introduce measurement error. You can still use the same procedure that we advocate without the switch, but you will have to transfer the voltmeter leads back and forth a few times.

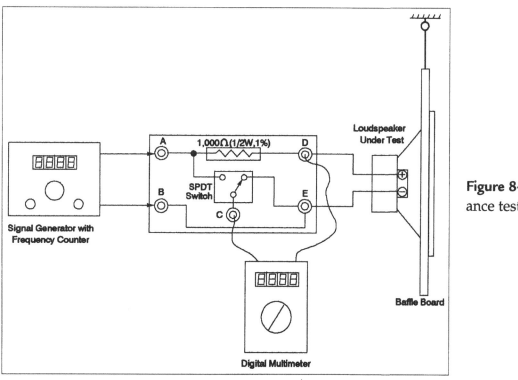

Figure 8-1 Impedance testing circuit.

The test equipment required is a signal generator, a frequency counter (preferably built into the signal generator), a high-impedance ac voltmeter, a 1,000-ohm, ½-W resistor (preferably 1 percent precision), and an assortment of 1 percent precision resistors used to check calibration. Referring to Figure 8-1, the signal generator is connected to input terminals A and B. The loudspeaker under test is connected to terminals D and E. The positive (red) lead from the ac voltmeter is connected to terminal D and the negative (black) lead is connected to terminal C. The single-pole, double-throw switch allows you to conveniently toggle the black lead back and forth between terminals A and E.

Calibration Procedure

1. In place of the loudspeaker, connect a precision resistor of known value across terminals D and E.
2. Set the signal generator for a sine-wave output and turn on the power. Set the frequency for an arbitrary 30 Hz.
3. Set the multimeter to read ac volts. With the switch set to terminal A, adjust the generator output to obtain a reading in millivolts which is numerically equal to the series resistor's value. For example, if the series resistor is 1,000 ohms, set the generator to read 1,000 mV or 1.0 V.
   Note: Allow approximately 15 minutes for equipment temperatures and voltages to stabilize before proceeding to the next step.
4. Flip the switch to terminal E. The millivolt reading that appears should be equal to the resistance of the test resistor in ohms. For example, if your test resistor is 10 ohms, the voltage reading should be 10 mV or 0.010 V. If the reading is different, adjust the signal generator output until you obtain the exact reading.
5. Without making any adjustments, reset the switch to terminal A and record the exact voltage. For example, if you are using a 1,000-ohm series resistor, the voltage may read 995 or 1,005 mV.
6. Repeat steps 4 and 5 with different calibration resistors ranging in value from approximately 4.7 to 100 ohms. After replacing a resistor, reset the switch to terminal E, and again adjust the generator output until you obtain a reading exactly equal to the test resistor value. Flip the switch to terminal A and record the exact millivolt reading.

7. When all of the test resistors have been used, take the average of the series resistor readings found in steps 5 and 6. (If your meter and test setup are working properly, there may be only small differences in these readings for the various calibration resistors.) Record this value on the test circuit for use on all future measurements. We will refer to this in the future as the "calibration value."

Determining Q_{TS}

Before you test a new woofer, you should always break it in first to loosen up the suspension. To do this, connect the woofer to the signal generator and apply a moderate signal of about 30 Hz. If your signal generator doesn't have sufficient output to produce visible cone movement, you will need to amplify the signal. Listen for any unusual noise that would indicate that the woofer is being over-driven. If you hear any, reduce the signal level. The woofer should be broken in for a minimum of an hour.

The general shape of a dynamic loudspeaker's impedance curve is shown in Figure 8-2. It has a resonance peak at frequency f_s and, after a relatively brief flat section, increases with frequency, primarily due to voice-coil inductance. The points of interest we will be initially concerned with are frequencies f_S, f_1, and f_2, R_M, and R_1, which define the shape of the resonance peak and therefore the total Q.

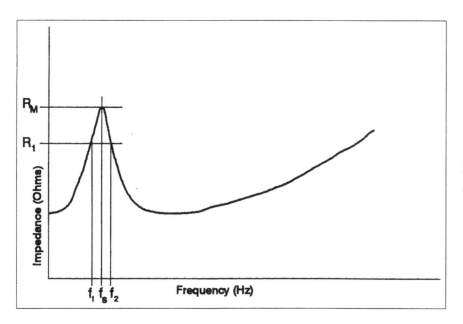

Figure 8-2 Typical loudspeaker impedance curve.

1. We begin first by measuring the woofer's voice-coil dc resistance. Record this value as R_E. Notes on measuring R_E:
   - Low resistance values are difficult to do accurately with a low-cost multimeter. If you are using one, you will need to verify its accuracy by testing it with one of the precision resistors with a value near the voice coil's advertised resistance. If the meter reading is the same as the calibration resistor, you can proceed with confidence.
   - For all of our example tests, we will be using the Radio Shack Catalog No. 22-174B because of its versatility, availability, and moderate cost. However, it does have its limitations. The sample I purchased indicated 1.2 ohms instead of zero (as did two others in the store) when I clipped the leads together on the resistance setting. This zero offset was verified with the 1 percent calibration resistors and my old standby Fluke Model 77. In order to use this meter for low-resistance measurement, we will have to subtract 1.2 ohms from the reading.
   - This meter will also measure frequency, but unfortunately not to within 0.1 Hz. So if you try to rely on it exclusively for low-frequency measurements, your results could be off by 5 to 10 percent.
2. Next, we must suspend the woofer vertically in free space by hanging it with a string and connecting it to the circuit shown in Figure 8-1. (To obtain measurements more representative of how the speaker will behave when mounted, you should test it mounted on a baffle board with approximately the same dimensions you plan to use.) (Reference 3.) With the switch set to E, adjust the signal generator frequency until the maximum ac voltage is read on the multimeter.
3. Flip the switch to A and adjust the signal generator output until you obtain the calibration reading we recorded in step 7 above.
4. Flip the switch back to E and sweep the frequency up and down to verify the maximum impedance point. Record the frequency where this maximum voltage occurs as f_s and the maximum impedance as R_M.
5. Calculate r_0:

$$r_0 = R_M / R_E \qquad (8\text{-}1)$$

6. Calculate R_1:

$$R_1 = R_E(r_0)^{1/2} \tag{8-2}$$

7. Referring to Figure 8-2, reduce the signal generator frequency to the point where the impedance is equal to R_1. To maintain accuracy, we will have to recalibrate at this lower impedance.
8. Flip the switch to A and reset the signal generator output until the meter reading is equal to the calibration value.
9. Flip the switch back to E and sweep the frequency carefully until the voltmeter again indicates the value of R_1. This should now be the accurate frequency location for R_1. Record the frequency at this point as f_1.

   Note: You can verify that the system is still in calibration by flipping the switch back to A. The voltmeter reading should show little if any difference. If it does, reset the generator output again to obtain the calibration value and repeat step 9.

10. Next, slowly increase the frequency until the impedance is again equal to R_1. When you are in the vicinity, be sure to recalibrate before taking your final reading. Record this frequency as f_2.
11. If your measurement system is accurate, f_s will equal $(f_1 f_2)^{1/2}$ within a few percent. If it differs by more than 1 or 2 Hz, you have excessive measurement error in your system, and you should attempt to determine the cause before proceeding.
12. Calculate Q_{MS} (the mechanical component of Q_{TS} measured in free air):

$$Q_{MS} = f_s(r_0)^{1/2} / (f_2 - f_1) \tag{8-3}$$

13. Calculate Q_{ES} (the electrical component of Q_{TS} measured in free air)

$$Q_{ES} = Q_{MS} / (r_0 - 1) \tag{8-4}$$

14. Calculate Q_{TS} (total Q measured in free air):

$$Q_{TS} = 1 / (1/Q_{MS} + 1/Q_{ES}) \tag{8-5}$$

Determining V_{AS}

The third Thiele-Small parameter V_{AS} is related to compliance. In the most simplistic terms, it can be thought of as the relative amount of air the driver will displace. The larger this number, the larger the enclosure the driver will require. There is universal agreement that it is difficult to measure V_{AS} accurately (Refs. 3, 5, 11), and three methods are given in the literature for measuring it. The first two require that you build a test box of known volume. The method, referred to as the "added mass method," is described in Ref. 3 and does not require a test box. Instead it involves carefully adding a mass of known weight to a speaker cone and measuring the difference between the free-air resonance frequency f_s with and without the mass. The accuracy of this method is mostly dependent on how well you estimate the effective piston diameter of the driver. On small drivers, the surround represents a relatively large percentage of the moving area and makes it more difficult to estimate the effective diameter accurately.

Method 1 appears to be the most straightforward and is referred to as "the closed-box method." It is more of a direct measurement of the "volume equivalent air spring" and is based on the change in resonant frequency associated with pumping a known volume of air in a closed box. Figure 8-3 gives some recommended test-box volumes for various driver sizes. To assure accuracy, the test box should cause a resonant frequency increase of at least 25 percent. To calculate V_{AS}, use the following equation (Refs. 3 and 11):

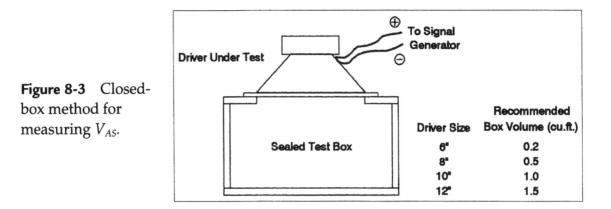

Figure 8-3 Closed-box method for measuring V_{AS}.

Driver Size	Recommended Box Volume (cu.ft.)
6"	0.2
8"	0.5
10"	1.0
12"	1.5

$$V_{AS} = V_T[(f_{CT}Q_{ET}/f_SQ_{ES}) - 1] \qquad (8\text{-}6)$$

where V_T = volume of the test box

$\quad f_{CT}$ = resonant frequency of the driver and test box

$\quad Q_{ET}$ = electrical Q of the driver and test box

Application

For our first example, we'll test a Madisound 5502R-4 that we will be using in an experimental two-way system. This is a 5¼-in polypropylene woofer with a nominal 4-ohm impedance.

For all testing, I mounted the driver in a door-panel-type baffle and teamed it with a Vifa D26NC-05-06 soft dome tweeter. This assembly was installed in the test wall (Photo 8-1). After the 1-hour break-in, I took the following measurements with the test circuit of Figure 8-1:

	Measured	Published specifications
R_E, ohms	3.6	3.64
f_S, Hz	54.9	48
Q_{MS}	3.72	2.89
Q_{ES}	0.53	0.47
Q_{TS}	0.46	0.40

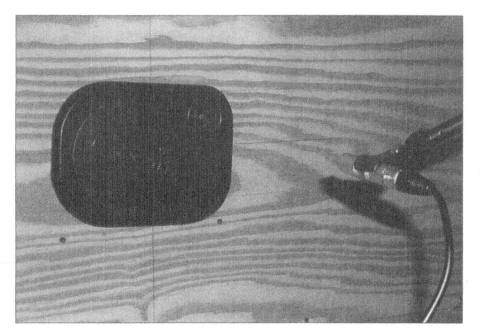

Photo 8-1

With the exception of R_E, all of the measured parameters were somewhat different from the published specifications but were within what I would consider to be normal manufacturing tolerances. The second sample measured similarly, so I have to conclude that the samples are representative of this particular manufacturing lot.

To measure V_{AS}, I installed the bass driver on a 375-in³ test box (Photo 8-2); the resonant frequency increased from 54.9 to 81.0 Hz and Q_{ES} increased from 0.527 to a closed-box Q_{EC} of 0.835. Using Eq. (8-6), V_{AS} calculates to be 8.22 liters, significantly less than the published value of 13.8 liters, which I might add, is not unexpected with the higher value of f_S.

Photo 8-2

With these parameters, *Spreakerbuilder* loudspeaker design software predicted a low-frequency cutoff (or –3-dB point) for the driver mounted on an infinite baffle in free air of 85 Hz, not exactly what I would call "bass," but good enough to be classified as "midbass." *Spreakerbuilder* is shareware you can download from AOL's library, but one of its shortcomings is that it won't print any of the response curves. On the other hand, it is very handy for calculating closed and ported boxes and passive crossover networks. Since *Spreakerbuilder* wouldn't print anything, I printed the predicted response using *LspCAD Lite 1.0*. This is shown as Figure 8-4a. Later on, we will give brief overviews of these and a few more shareware programs. Being shareware, virtually none of them are what I would call complete; however, each has some standout features.

Figure 8-4*b* shows the predicted response for our test woofer in a 0.49-ft³ ported box with a box frequency of 44.5 Hz. The –3-dB point is 39.5 Hz, which is outstanding for a 5¼-in driver. Given the proper environment, this driver should perform extremely well.

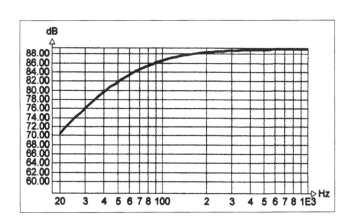

Figure 8-4*a* Predicted frequency response of sample Madisound 5502R-4 using LspCAD 1.0 by Ingemar Johansson. SPL was calculated for a distance of 1 m with an input of 2.83 V. Mounted on infinite baffle in free air.

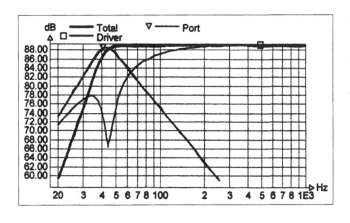

Figure 8-4*b* Predicted frequency response of sample Madisound 5502R-4 using LspCAD 1.0 by Ingemar Johansson. SPL was calculated for a distance of 1 m with an input of 2.83 V. Installed in 0.49-ft³ vented box tuned to 44.5 Hz.

Two-Way Crossover Design

Designing passive crossovers is, without a doubt, the most difficult job for amateur speakerbuilders. With modern computers and software, this process has been made much easier, but those avenues are still fairly expensive. With some trepidation, I will attempt to present here an economical methodology that seems to work fairly well. Frankly, there will be nothing new or different in this process other than it will be presented from a hands-on perspective.

The crossover design process is much simpler than it would appear. In summary the steps are:

1. Measure the bass driver's Thiele-Small parameters F_S, Q_{TS}, and V_{AS}.
2. Design the impedance compensation network for the woofers.
3. Mount the drivers on a common baffle.
4. Measure the relative sensitivities of the drivers at the crossover frequency and at selected frequencies above and below it.
5. Measure the impedance of the drivers at the frequencies selected in step 5.
6. Calculate crossover network component values.

If you are mounting the woofer in free air (on a door panel), you can even omit step 1 as it has no influence on the rest of the crossover design process.

Procedure

1. Solder several feet of lead wire to each of the drivers to be tested.
2. Mount the drivers on a large baffle or install them in the cabinet you plan to use. If using a cabinet, run all of the lead wires outside the cabinet through sealed openings and place the cabinet on a stand 2 to 3 ft above the floor. (While this is not an anechoic measurement, it may actually be more representative of room response.)
3. The first step in the crossover design process is to design the woofer's impedance compensation network. Looking back to Figure 8-2, we noted earlier the rise in the impedance curve associated with voice coil inductance. If this is not equalized, the crossover will not work properly—the woofer will exhibit a similar rising response in the vicinity of the crossover point. We can compensate for this by placing a resistor and capacitor in series across the speaker terminals, shown in Figure 8-5

as R_Z and C_Z. This is also referred to as a "Zobel" network. There are some rules for calculating the values of R_Z and C_Z that will get you in the ballpark, but you will have to do a little experimentation to get the flattest impedance. Reference 3 recommends that the starting point is to set R_Z equal to 1.25 times R_E, the voice coil dc resistance. Another author recommends setting it equal to the driver's nominal impedance—in this case 4 ohms. The author of LspCAD design software sets it equal to R_E. Taking all that into consideration, I had some 4-ohm resistors, so I used one of them.

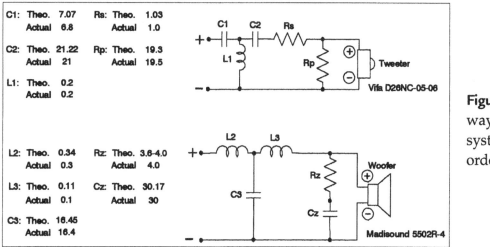

C1:	Theo.	7.07	Rs:	Theo.	1.03
	Actual	6.8		Actual	1.0
C2:	Theo.	21.22	Rp:	Theo.	19.3
	Actual	21		Actual	19.5
L1:	Theo.	0.2			
	Actual	0.2			
L2:	Theo.	0.34	Rz:	Theo.	3.6–4.0
	Actual	0.3		Actual	4.0
L3:	Theo.	0.11	Cz:	Theo.	30.17
	Actual	0.1		Actual	30
C3:	Theo.	16.45			
	Actual	16.4			

Figure 8-5 Two-way experimental system with third-order crossover.

Reference 3 also provides a calculation method to determine the value of C_Z based on the voice coil inductance. However, it is fairly easy to find by trial and error if you have an assortment of nonpolarized capacitors lying around. Since I happened to have LspCAD 3.10 on my PC and the published voice coil inductance value, I ran the numbers through and came up with a value of 30 μf. Placing these across the woofer terminals, I ran an impedance check by sweeping the frequencies up and down out to 6,000 Hz and found it to be flat within about 5 percent.

4. With the woofer impedance equalized, we are ready to measure the sensitivities of each of the components relative to each other at and around the crossover frequency we have chosen (well in advance) for the system.

Notes:

- For a two-way system, you will obtain wider dispersion, and therefore a wider soundstage, by crossing over to the tweeter at the lowest frequency possible. This is limited by the capability of the tweeters. If you set the crosspoint too low, they can be easily damaged. The best way to avoid this is to adhere to the guidelines provided by the tweeter manufacturer. As a minimum, the crossover point should be double the tweeter's free-air resonance frequency.

- Increasing the crossover "order" will enable you to set the crosspoint lower, since the steeper cutoff blocks more of the low-frequency content. Noted authority G. R. Koonce, contributing editor to *Speakerbuilder* magazine, has recommended third-order crossovers for their predictability and practicality (Ref. 2). Based on his recommendation and my own personal experience, I selected a third-order crossover for our example design.

- By far the cheapest way to measure sound pressure levels at moderate frequencies is with the Radio Shack sound-level meter. To use it properly, it should be mounted on a boom-type stand. You can either make one with a few pieces of small lumber or purchase one from a music store for $30 or $40.

- Position the sound-level meter midway between the woofer and tweeter, similar to the microphone placement shown in Photo 8-1. To minimize room effects, set the distance at about 18 in for a two-way system. You'll have to double that for a three-way system, so you may as well make it a full meter (39.37 in) to put things on the same basis as published specifications. At that distance, however, the room will have a much larger effect on your measurements.

- To generate test signals, you will need a signal generator, an ac voltmeter capable of measuring voltages at several thousand hertz (like the Radio Shack 22-174B), and some type of amplifier. The test setup is shown in Figure 8-6.

To measure the relative sensitivity of the drivers, set the signal generator frequency to the crossover frequency and adjust the output level to the industry standard of 2.83 V. Connect the woofer to the amplifier and readjust the signal

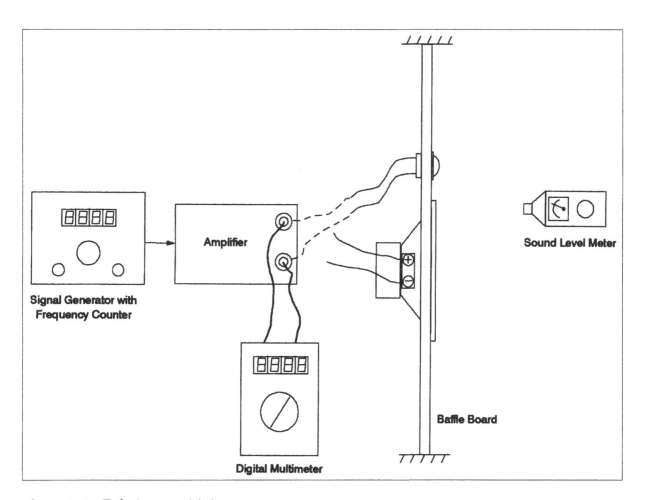

Figure 8-6   Relative sensitivity test setup.

generator output to 2.83 V. Read the sound-level meter by observing from a distance as far away as practical to avoid influencing the measurement and record the SPL for the woofer. Without making any adjustments, disconnect the woofer and connect the tweeter. Check the amplifier voltage and readjust the signal generator output to maintain the amplifier voltage of 2.83 V. Now record the sound-level meter reading for the tweeter.

Lower the signal generator frequency setting 500 Hz below the crossover point and repeat the above measurements. Do this again for 1,000 Hz above the crossover point.

Now, average the sets of readings and compare the difference and record this number for future reference. If you selected the right components, the tweeter will have at least the same sensitivity as the woofer. Typically, it will be 2 to 4 dB higher. If the tweeter has a lower sensitivity than the woofer, then the two are mismatched.

5. Connect the impedance measuring circuit of Figure 8-1 and carefully measure the impedance at each frequency/sensitivity measurement point. Set the signal generator at the frequency point, connect the woofer, calibrate the impedance measurement circuit, and measure and record the impedance. Disconnect the woofer and connect the tweeter without disturbing the frequency setting. Recalibrate and measure the tweeter's impedance.

For my two sample drivers, I made the following measurements after installing the woofer Zobel:

	Woofer		Tweeter	
Frequency (Hz)	SPL (dB)	Impedance (Ω)	SPL (dB)	Impedance (Ω)
2,500	96.2	4.3	98.8	5.13
3,000	96.3	4.3	104.5	4.98
4,000	100.0	4.2	97.0	4.68
Average	97.5	4.3	100.1	4.93

6. Up until now, everything else in the crossover design process has been focused on trying to characterize the actual driver parameters. We have determined the relative difference in sensitivities between our drivers and we know their average impedances in the vicinity of the crossover point. Calculation of the crossover component values is almost trivial in comparison. You can use design formulas available in Refs. 3, 8, 9, and 10, or use your PC and any of the shareware programs. Both Speakerbuilder and Blaubox will calculate them by inputting the desired crossover frequency and driver impedance at that frequency. I chose 4.3 ohms for the woofer impedance and 5.0 ohms for the tweeter and got the values shown in Figure 8-5 for a 3,000-Hz crossover point.

The "breadboarded" crossover circuits and drivers are shown in Photo 8-3.

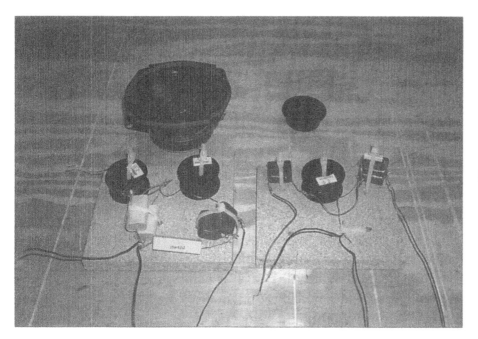

Photo 8-3

Measured Response

Finally, the moment of truth.... Figure 8-7*a* is the measured response of the system in the test wall with the mike positioned midway between the woofer and tweeter at a distance of 18 in. With the exception of the single dip of −6 dB at 250 Hz, everything else is ±3 dB from 80 Hz to 20 kHz. However, looking at the tweeter response, there are more points above the 0-dB line than below it, which is not really a surprise, since the sensitivity data averages indicate that the tweeter is 2.6 dB "hotter" than the woofer.

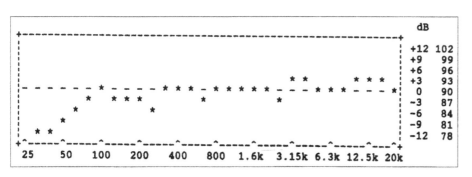

Figure 8-7*a* Measured frequency response Madisound 5502R-4 and Vifa D26NC-05-06 two-way system with third-order crossover network. No tweeter L-pad.

Figure 8-7b Measured frequency response Madisound 552R-4 and Vifa D26NC-05-06 two-way system with third-order crossover network. With tweeter L-pad.

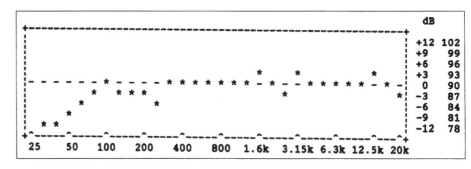

One way to reduce tweeter output is to insert a series resistor. However, if we do that and nothing else, the impedance seen by the tweeter crossover would increase, which would alter the tweeter's crossover frequency, which is not good. We need a way to lower the tweeter output without changing the impedance load seen by the crossover network. This calls for one more tool from our speaker-builder toolbox to bring this project to completion, and that tool is the L-pad attenuator.

The L-pad consists of two resistors, one in series with the driver R_P, which does the attenuation, and R_P in parallel with the driver, which maintains the load seen by the crossover after inserting R_S.

Circuit Analysis

Referring to Figure 8-8, R_S and R_P in parallel with the tweeter (with impedance Z_D) form a simple voltage-divider circuit shown in the equivalent circuit to the right. For this circuit:

Figure 8-8 Tweeter L-pad and equivalent circuit.

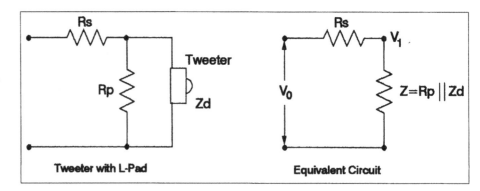

$$V_1/V_0 = [Z/(R_S + Z)] \tag{8-7}$$

where

$$Z = R_P \parallel Z_D = (1/R_P + 1/Z_D)^{-1} \tag{8-8}$$

The ratio of V_1 to V_0 is the attenuation A. In logarithmic decibel terms this is expressed as

$$20 \log V_1/V_0 = A \qquad \text{(in dB)} \tag{8-9}$$

Then

$$V_1/V_0 = 10^{A/20} \tag{8-10}$$

Substituting (8-10) into (8-7),

$$10^{A/20} = [Z/(R_S + Z)] \tag{8-11}$$

For the load on the crossover circuit to remain constant,

$$R_S + Z = Z_D \tag{8-12}$$

(the tweeter's impedance before adding R_S). Substituting into (8-11),

$$10^{A/20} = Z/Z_D = (1/R_P + 1/Z_D)^{-1}/Z_D \tag{8-13}$$

Solving for R_P,

$$R_P = Z_D(10^{A/20})/(1 - 10^{A/20}) \tag{8-14}$$

Rearranging (8-12),

$$R_S = Z_D - Z$$

$$R_S = Z_D - (1/R_P + 1/Z_D)^{-1} \tag{8-15}$$

As a first trial, I decided to try −2 dB for the tweeter's attenuation. Substituting this for A and the tweeter's impedance of 5 ohms for Z_D in Eqs. (8-8) and (8-9) yields

$$R_P = 19.3 \text{ ohms} \qquad \text{and} \qquad R_S = 1.03 \text{ ohms}$$

Actual values used for R_P and R_S were 19.5 and 1.0 ohms, respectively.

After installing the L-pad in the tweeter circuit, I remeasured the system response, which is shown as Figure 8-7b. We still have a small anomaly at the crossover point which I'm confident could be made seamless with a little tinkering. However, that is not the purpose of this experiment. "Right out of the box" this system has a response that is flat within ±3 dB, and much of it is ±1.5 dB. In general, it sounds quite smooth, as I would expect from the response plot.

Having the ability to actually measure the response of the system is greatly satisfying, to say the least. It also gives us the means to make small adjustments that improve the overall smoothness, making it more listenable and less fatiguing. The Audio Control SA-3055 real-time analyzer, because of its accuracy and ease of use, makes this part both fun and convenient. However, as they say, it's not cheap. If your budget won't stand that kind of investment, you can trade convenience for more time by using a test CD which contains bands of filtered pink noise and the Radio Shack sound-level meter. The sound level meter's frequency response is limited to 20 to about 8 kHz. But that's enough bandwidth to check out almost any passive crossover network. A suggested test CD is *My Disk* from Sheffield Labs and Autosound 2000.

Now the point of all this was to demonstrate that we can, as amateurs, design passive crossovers that work fairly well using the standard tools that are available to us in the literature. Of course, we still have to invest a few hundred dollars in test equipment, but this is the kind of thing that can bring a lifetime of enjoyment by enabling us to build speakers for both the car and home.

Some Observations on Speaker Design Shareware

As stated earlier, none of the shareware programs are what I would call complete, and in fact most of them are intended to give you just enough features during the test drive to entice you to pay the registration fee and get the rest of the program. The exception here is *Blaubox*, which is now freeware. So let's take a brief look at a few of my favorites.

Speakerbuilder

- Windows 3.1 based.
- Available for downloading from AOL's Windows software library.
- Has driver database for storing Thiele-Small parameter files.
- Calculates theoretical closed and vented box alignments.
- Plots bass box responses, up to three simultaneous plots for comparison.
- Allows user input of nonoptimal parameters for comparison.
- Has extremely handy box and port dimension calculation screen, including mid-tweeter–mid-D'Appolito driver spacing.
- Perhaps its best feature is its passive crossover design utility. While it only does theoretical, ideal networks, it will handle anything from first-order two-ways to fourth-order three-ways, and anything in between with your choice of seven different roll-offs, from maximally flat Butterworths to Chebychevs with a Q of 1. Biggest shortcoming—the shareware version won't print anything.

Loudspeaker CAD Lite 1.0 by Ingemar Johansson

- Windows 3.1 based.
- Can be downloaded at www.caa.pair.com
- Considered by many to be the best of the shareware programs.
- Comes with large preloaded driver database.
- Cross-calculates driver physical and electrical parameters.
- Calculates bass boxes for closed, vented (including percent fill), and fourth-order bandpass. Has powerful bandpass design capabilities.

- Design screen allows alteration of any of the design parameters with instantaneous screen update.
- Plots and prints frequency response, driver impedance, cone excursion, and port airspeed. Also shows port standing-wave modes.
- Sample response plots are shown in Figures 8-4*a* and *b*.

For a modest $30, an even more powerful version (3.10) is available via Internet from its author which will do passive radiators, both fourth- and sixth-order bandpass boxes, and two- and three-way crossover calculations, including both woofer impedance equalization and tweeter L-pads. An interesting feature is that it will design a passive crossover with little more than the drivers' voice coil dc resistance and inductance. The full commercial version is listed for $129.

Perfect 4.5 by Warren Merkel

This program is based on Bullock and White's *Boxresponse* program and has several unique features. It has been licensed in the past by MTX under the name *Thunderbox* and was quite expensive.

- DOS based.
- Can be downloaded at www.caa.pair.com or www.diyloudspeakers.org
- Comes with a huge (850) driver base (probably most of which is obsolete by now).
- Calculates closed and ported boxes.
- Plots frequency response and lists −3-dB point and the amount of response ripple.
- Unique to this group, it includes calculated responses with the active second-order equalizer that is also part of my subwoofer filter and will even show the schematic.
- Plots cone excursion and SPL versus frequency.
- Plots up to four simultaneous response curves.
- Will print data points used in graphs.
- A sample response curve of a Thiele-Small sixth-order alignment is shown as Figure 8-9.

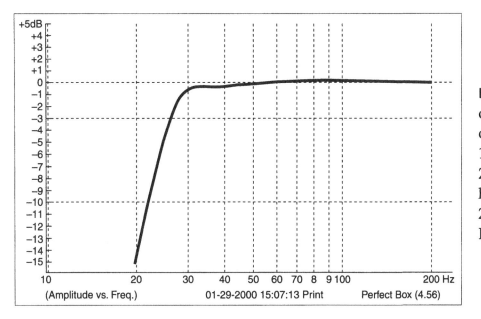

Figure 8-9 Sample of a perfect 4.5 printout. Madisound 10207; 3-dB point is 21 Hz with a 0.27-dB hump. V_b = 1 ft^3. F_b = 21 Hz. Order = 6. EQ at 27 Hz. D= 0.4.

Blaubox by Blaupunkt

While not the most user-friendly in this group, *Blaubox* has quite a few capabilities.

- DOS based.
- Can be downloaded at www.caa.pair.com or www.diyloud-speakers.org
- Comes preloaded with a small driver database, which you add to.
- Calculates closed, vented, and fourth-, fifth- (fourth with a series inductor), and sixth-order bandpass enclosures for both regular and isobaric systems.
- Plots and prints two simultaneous frequency responses.
- Has a box fabrication feature to aid in actual panel sizing and will actually print out a box drawing.
- Calculates ideal first-, second-, and third-order two-way passive crossovers.
- Best feature is bandpass design.
- A sample response curve is shown in Figure 8-10.

Figure 8-10 Sample response printout from Blaubox. Loudspeaker design software by Blaupunkt. Sample Madisound 5502R-4 5¼-in woofer mounted on infinite baffle.

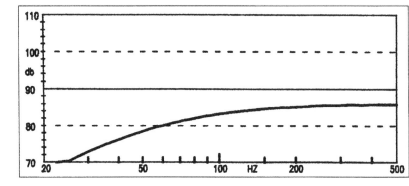

My Spreadsheets

To be able to predict how my subwoofer filter will affect a bass enclosure/driver combination, I have combined the transfer function for the filter with those for closed and vented boxes given in Ref. 20 into two separate spreadsheets. You can reproduce them by entering the cell listings below.

For Vented Box Thiele-Small Sixth-Order Alignments:

Row 1 is the title. Here, I usually enter the driver model and the type alignment.

Row 2 is a dividing line.

Row 3 is the list of frequencies to be plotted, which you can modify as you see fit.

For the frequencies I have chosen:

A3: 'Frequency, B3: 10, C3: 15, D3: 20, E3: 25, F3: 28, G3: 30, H3: 35, I3: 40, J3: 45,

K3: 50, L3: 55, M3: 60, N3: 65, O3: 70, P3: 75, Q3: 80, R3: 85, S3: 90, T3: 95, U3: 100,

V3: 120, W3: 140, X3: 150, Y3: 180, Z3: 200, AA3: 250, AB3: 300, AC3: 350, AD3: 400

Row 4 is a dividing line.

A5: 'Qts, B5: Enter the driver's Qts, C5: +B5, Copy C5 across to AD5

A6: 'Fs, B6: Enter the driver's Fs, C6: +B6, Copy C6 across to AD6

A7: 'Vas, B7: Enter the driver's Vas, C7: +B7, Copy C7 across to AD7

A8: 'Vb, B8: Enter the box volume in cubic feet, C8: +B8, Copy C8 across to AD8

A9: 'Qb, B9: Enter the box loss factor, C9: +B9, Copy C9 across to AD9

Note: Box loss factor can range from 3 to 15. Typical value is 7.

A10: 'fb, B10: Enter the box frequency, C10: +B10, Copy C10 across to AD10

A11: 'Alpha, B11: +B7/B8, Copy B11 across to AD11

A12: 'h, B12: +B10/B6, Copy B12 across to AD12

A13: 'fn, B13: +B3/B6, Copy B13 across to AD13

A14: 'A, B14: +B12^2, Copy B14 across to AD14

A15: 'B, B15: +B14/B5+B10/(B9*B6), Copy B15 across to AD15

A16: 'C, B16: 1+B14+B11+B10/(B6*B9*B5), Copy B16 across to AD16

A17: 'D, B17: 1/B5+B10/(B6*B9), Copy B17 across to AD17

Row 18 is a dividing line.

A19: 'VB Resp,

B19: 20*@Log(B13^4/((B13^4-B16*B13^2+B14)^2+B13^2*(B17*B13^2-B15)^2)^0.5)

Copy B19 across to AD19

Note: Row 19 is the unassisted vented box response.

Row 20 is a dividing line.

A21: 'X Freq, B21: +B3, D21: +D3, G21: +G3, K21: +K3, O21: +O3, R21: +R3,

U21: +U3, X21: +X3, Z21: +Z3, AB21: +AB3, AD21: +AD3

Note: Row 21 entries are the X-axis frequency markers for the graphing function.

Row 22—Copy Row 3 here to repeat frequency numbers.

Row 23: Another dividing line.

A24: 'High-Pass Filter Section

A25: 'QfHP, B25: Enter the high-pass filter Q, C25: +B25, Copy C25 across to AD25

A26: 'FfHP, B26: Enter the high-pass filter characteristic frequency, C26: +B26

Copy C26 across to AD26

A27: 'HP Resp, B27: 20*@Log(B22^2/((B26^2-B22^2)^2+(B26*B22/B25)^2)^0.5)

Copy B27 across to AD27

Row 28 is a dividing line.

A29: 'Low-Pass Filter Section

A30: 'Order 1

A31: 'FfLP1, B31: 159, C31: +B31, Copy C31 across to AD31

A32: 'LPResp1, B32: 20*@Log(B31/(B31^2+B22^2)^0.5)

A33:'Order 2

A34: 'FfLP2, B34: Enter the subwoofer crossover frequency (30 to 330 Hz), C34: +B34

Copy C34 across to AD34

A35: 'QfLP2, B35: 1, C35: +B35, Copy C35 across to AD35

A36: 'LPResp2, B36: 20*@Log(B34^2/((B34^2-B22^2)^2+(B34*B22/B35)^2)^0.5)

Copy B36 across to AD36

A37: 'LPRespTot, B37: +B32+B36, Copy B37 across to AD37

Row 38 is a dividing line.

A39: 'FiltResp, B29: +B27+B37, Copy B29 across to AD29

Row 40 is a dividing line.

A41: 'SystResp, B41: +B19+B39

Row 42 is a dividing line.

A43: 'Vent Calculation

A44: 'Lv = 1.463*10^7*r^2/(Fb^2*Vb) − 1.463*r (a statement of the formula)

Note: Lv = Vent length in inches, r = Vent radius in inches, Fb = box frequency

from B10, Vb = Box Volume from B8 converted to cubic inches

A45: 'r, B45: Enter the vent radius in inches

A46: 'Fb, B46: +B10

A47: 'Vb, B47: +B8*1728

A48: 'Lv, B48: 1.463*10^7*B45^2/(B46^2*B47)-1.463*B45

Row 49 is a dividing line.

A50: 'High-Pass Filter Resistor Calculation (Calculates R1 and R2 for given filter Q and Frequency)

A51: 'Qfhp = 0.5*(R1/R2)^0.5 (A51 and A52 are statements of the basic equations).

A52: 'Ff = $1/(2*pi*C*(R1/R2)^{0.5})$

A53: 'Qfhp, B53: +B25

A54: 'R1/R2, B54: $(2*B53)^2$

A55: 'C, B55: 2.2E-07

   Note: C is the value of capacitor C7 and C8 used in the filter.

A56: 'Ff, B56: +B26

A57: 'R1, B57: $1/(2*3.1416*B55*B56*2*B53)$

A58: 'R2, B58: +B54*B57

For the Electronically Assisted Closed-Box Alignments

Row 1 is the title. Again, enter the driver model and the type alignment.

Row 2 is a dividing line.

Row 3 is the list of frequencies to be plotted, which (again) you can modify as you see fit.

For the frequencies I have chosen:

A3: 'Frequency, B3: 10, C3: 15, D3: 20, E3: 25, F3: 28, G3: 30, H3: 35, I3: 40, J3: 45,

K3: 50, L3: 55, M3: 60, N3: 65, O3: 70, P3: 75, Q3: 80, R3: 85, S3: 90, T3: 95, U3: 100,

V3: 120, W3: 140, X3: 150, Y3: 180, Z3: 200, AA3: 250, AB3: 300, AC3: 350, AD3: 400

Row 4 is a dividing line.

A5: 'Qts, B5: Enter the driver's Qts, C5: +B5, Copy C5 across to AD5

A6: 'Fs, B6: Enter the driver's Fs, C6: +B6, Copy C6 across to AD6

A7: 'Vas, B7: Enter the driver's Vas, C7: +B7, Copy C7 across to AD7

A8: 'Vb, B8: Enter the actual box volume in cubic feet, C8: +B8, Copy C8 across to AD8

A9: 'Alpha, B9: +B7/B8, Copy B9 across to AD9

A10: 'Qtc, B10: $(B9+1)^{0.5}*B5$, Copy B10 across to AD10

A11: 'Fc, B11: +B6(B10/B5), Copy B11 across to AD11

Row 12 is a dividing line.

A13: 'CB Resp, B13: $20*@log(B3^2/((B11^2-B3^2)^2+(B11*B3/B10)^2)^{0.5})$

Copy B13 across to AD13

Row 14 is a dividing line.

A15: 'High-Pass Filter Section

A16: 'Ff, B16: Enter the HP filter characteristic frequency, C16: +B16, Copy C16 across to AD16

A17: 'QfHP, B17: Enter HP filter Q, C17: +B17, Copy C17 across to AD17

A18: 'HP Resp, B18: $20*@Log(B3^2/((B16^2-B3^2)^2+(B16*B3/B17)^2)^{0.5})$, Copy B18 across to AD18

Row 19 is a dividing line.

A20: 'CB + HP Resp, B20: +B13+B18, Copy B20 across to AD20

Row 21 is a dividing line.

Row 22 is a duplicate of Row 3.

A23: 'X Freq, D23: +D22, I23: +I22, L23: +L22, P23: +P22, T23: +T22, Y23: +Y22,

AD23: +AD22

Row 24 is a dividing line.

A25: 'Low-Pass Filter Section

A26: 'First Order Section

A27: 'Ff, B27: 159, C27: +B27, Copy C27 across to AD27

A28: '1stOrdResp, B28: $20*@Log(B27/(B27^2+B22^2)^{0.5})$, Copy B28 across to AD28

Row 29 is a space.

A30: 'Second Order Section

A31: 'Ff, B31: Enter subwoofer crossover frequency (30 to 330 Hz), C31: +B31,

Copy C31 across to AD31

A32: 'QfLP, B32: 1, C32: +B32, Copy C32 across to AD32

A33: 'SecOrdResp, B33: $20*@log(B31^2/((B31^2-B22^2)^2+(B31*B22/B32)^2)^{0.5})$, Copy B33 across to AD33

Row 34 is a dividing line.

A35: 'Filter Resp, B35: +B18+B28+B33, Copy B35 across to AD35

Row 36 is a dividing line.

A37: 'Total Resp, B37: +B13+B35

Row 38 is a line.

A39: 'High-Pass Filter Resistor Calculation

A40: 'Qf = 0.5*(R1/R2)^0.5   (A40 and A41 state the basic equations)

A41: 'Ff = 1/(2*pi*(R1*R2)^0.5)

A42: 'Qf, B42: +B17

A43: 'R1/R2, B43: 4*(B42)^2

A44: 'C, B44: 2.2E-07

A45: 'Ff, B45: +B16

A46: 'R1, B46: 1/(2*3.1416*B44*B45*2*B42)

A47: 'R2, B47: +B43*B46

These spreadsheets were used to generate all of the alignments and response curves in Chapter 6 and elsewhere throughout the book. To use them, you simply plug in the values for the driver's Thiele-Small parameters, set the box volume and box frequency (for vented boxes), and enter some hypothetical filter settings. With a few iterations, you can quickly home in on the best alignment for any given situation. Perhaps the greatest value these spreadsheets provide is to make it easy to obtain the best possible response from a box that is significantly smaller than ideal. With that in mind, I will demonstrate how they work with one final example.

Analysis of Example Thiele-Small Sixth-Order Alignment System

Looking back to Figure 3-3 in Chapter 3, we constructed and tested a 1-ft^3 (net) ported subwoofer system that employed two Madisound 8154 8-in woofers. I carefully measured their Thiele-Small parameters using the method outlined above and got the following results:

	Sample A	*Sample B*	*Published*
R_E	4.7	4.6	4.55
f_S	32.6	34.3	30.6
Q_{MS}	12.2	12.3	9.7
Q_{ES}	0.24	0.24	0.28
Q_{TS}	0.23	0.24	0.27
V_{AS} (*Ltrs*)	45.8	44.6	49.5

As a check on accuracy, $(f_1f_2)^{1/2}$ was within ½ percent of f_S for both samples.

After installing the drivers back in the cabinet, I measured the box frequency using two different methods. First, I located the point of minimum impedance in the valley between the two peaks shown in Figure 8-13, and second I placed the sound-level meter in close proximity to one of the woofer cones and located the frequency where cone movement was at a minimum. Both methods indicated 31.6 Hz as the box frequency.

I averaged all of the measured parameter values, plugged them into the vented box spreadsheet, and generated the predicted response curves shown as Figure 8-11a. These were plotted with Freelance 4.0 to enhance their appearance.

Figure 8-11a Two Madisound 8154 8-in woofers in a 1-ft³ vented box. Measured box frequency = 31.6 Hz. Predicted response with measured response superimposed. * = measured response unfiltered. # = measured response with subwoofer filter. Fb = 31.6 Hz. Qf = 2.5 at 30 Hz. R1 = 4.82 K. R2 = 120.6K. F3 = 30 Hz.

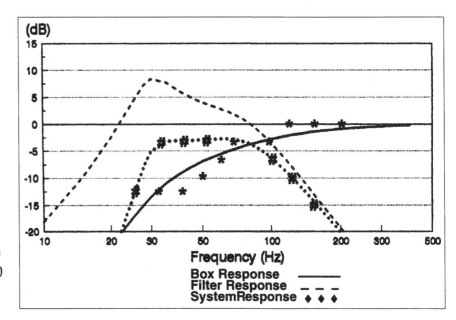

Figure 8-11b Two Madisound 8154 8-in woofers in a 1-ft³ vented box. Measured box frequency = 31.6 Hz. System response measured in garage with no filter.

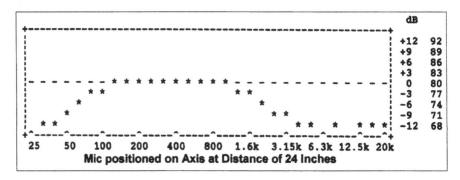

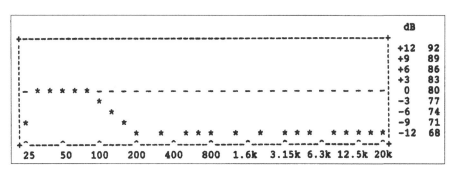

Figure 8-11*c* Two Madisound 8154 8-in woofers in a 1-ft³ vented box. Measured box frequency = 31.6 Hz. System response measured in a garage with subwoofer filter.

Next, I attempted to measure the unfiltered free-air response of the system outdoors, but got too much wind noise, so I retreated to the garage. With the door closed, I measured the system response with the cabinet lying horizontal at a height of 30 in off the floor. The mike was positioned on axis with the woofer nearest the port at a distance of 24 in. The result is shown in Figure 8-11*b*. I manually superimposed these points on the predicted response in Figure 8-11*a*. The correlation is only fair and tends to make the measurement technique suspect. Having said that, the filtered response correlation appears to be dead on. The measured response with the subwoofer filter in the circuit is shown in Figure 8-11*c*. When superimposed on Figure 8-11*a*, the curves appear identical.

When the system is placed on the back seat of a sedan, the response changes significantly. The unfiltered system response is shown as Figure 8-12*a*. While it appears to be quite rough, it has the characteristic low-frequency extension, which is well documented in the literature and seen in our past experiments. We again set the filter for flat response (no boost) and a 30-Hz cutoff and got the response shown in Figure 8-12*b*, which happens to be quite good.

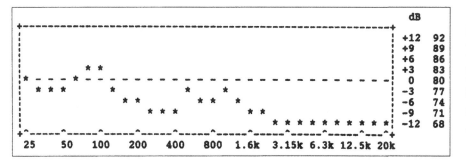

Figure 8-12*a* Madisound 8154 ported subwoofer system measured in-car response. Unfiltered response.

Figure 8-12*b*
Madisound 8154
ported subwoofer
system measured in-
car response. With
subwoofer filter set
for flat response and
30-Hz cutoff.

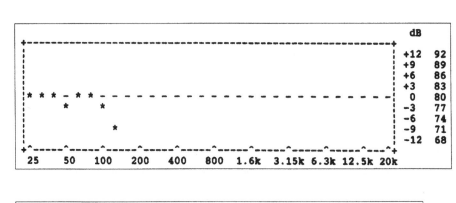

Figure 8-13 Pre-
dicted impedance
curve for Figure 8-11
system using
LspCAD Lite 3.10.

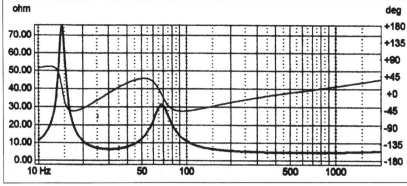

The point of this exercise was to demonstrate how to apply the spreadsheets. While they are not intended to take in-car response into account, they can still help you design a system that can be easily adjusted with the filter to adapt to nearly any environment. They are especially helpful for home applications, which more closely resemble free air.

Conclusion

That about wraps things up for this chapter and the book. My intention throughout all of this was to give you some new ideas on car stereo loudspeaker design that will enable you to build better and unique systems for any type of vehicle. In the process you should learn a lot more than if you simply paid someone else to do it for you. The final result should be more to your liking, and you will have the pride of knowing you did it yourself. If the system you build sounds as good as mine have, you'll never get tired of it and you'll be constantly thinking of ways to make it sound better.

Good luck with your system!

References

1. D'Appolito, Joe, "High-Pass Filter," SB Mailbox, *Speakerbuilder*, 2/90, p. 3.
2. Koonce, G. R., "Crossovers for the Novice," *Speakerbuilder*, 5/90, pp. 26–28, 30, 36, 38, 41–45, 91.
3. Dickason, V., *The Loudspeaker Design Cookbook*, Marshall Jones Co., 1988.
4. Bullock, R. M., "Thiele, Small, and Vented Loudspeaker Design, Part I," *Speakerbuilder*, 4/80, pp. 7–13, 30.
5. Bullock, R. M., "How You Can Determine Design Parameters for Your Loudspeakers," *Speakerbuilder*, 1/81, pp. 12–18.
6. Bullock, R. M., "Fine Points of Vented Speaker Design," *Speakerbuilder*, 2/81, pp. 18–25, 31, 35.
7. Bullock, R. M., "Thiele, Small, and Vented Loudspeaker Design Part V: Sixth Order Alignments," *Speakerbuilder*, 1/82, pp. 20–24.
8. Bullock, R. M., "Passive Crossover Networks: Part I," *Speakerbuilder*, 1/85, pp. 13–23.
9. Bullock, R. M., "Passive Crossover Networks: Part II," *Speakerbuilder*, 2/85, pp. 26–39.
10. Bullock, R. M., "Passive Crossover Networks: Part III," *Speakerbuilder*, 3/85, pp. 14–19.
11. Weems, D. B., *How to Design, Build, and Test Completed Speaker Systems*, Tab Books, 1978.

12. Jung, W. G., *Audio IC Op-Amp Applications*, Howard W. Sams & Co., 1981.

13. Berlin, H. M., *Design of Active Filters with Experiments*, Howard W. Sams & Co., 1981.

14. Knittle, Max R., "Microcomputer Driver Attenuation," *Speakerbuilder*, 1/85, pp. 24–25.

15. Armington, Richard, *Speaker Builder 1.0*, Windows-based shareware, available on America On Line.

16. *Blaubox*, loudspeaker design software by Blaupunkt.

17. Merkle, Warren, *Perfect 4.5*, loudspeaker design software.

18. Johansson, Ingemar, *LspCAD Lite*, Version 1.0, loudspeaker design software.

19. Johansson, Ingemar, *LspCAD Lite*, Version 3.1, loudspeaker design software.

20. Koonce, G. R., "Computing Box Responses," *Speakerbuilder*, 3/91, pp. 88–89.

21. Ferguson, D. L., *Ultimate Auto Sound*, Audio Amateur Press, 1995.

Resources

1. Madisound Speaker Components
   8608 University Green
   P.O. Box 44283
   Madison, WI 53744-4283
   Phone: 608-831-3433
   Fax: 608-831-3771

2. Parts Express
   340 E. First Street
   Dayton, OH 45402-1257
   Phone: 937-222-0173

3. Trendlines
   135 American Legion Highway
   Revere, MA 02151
   Phone: 800-767-9999

4. Woodworker's Supply
   1108 North Glenn Road
   Casper, WY 82601
   Phone: 800-645-9292

5. Audio Control
   22410 70th Avenue West
   Mountlake Terrace, WA 98043-2182
   Phone: 425-775-8461
   Fax: 425-778-3166
   www.audiocontrol.com

6. Old Colony Sound Lab
   305 Union Street
   P.O. Box 876
   Peterborough, NH 03458-0876
   Phone: 603-924-6371

7. Crutchfield
   1 Crutchfield Park
   Charlottesville, VA 22906
   Phone: 800-955-3000

8. MCM Electronics
   650 Congress Park Drive
   Centerville, OH 45459-4072
   Phone: 800-543-4330

9. The Auto Sound Lab
   782 Springbrook Lane
   Evans, GA 30809
   Phone: 706-860-4350
   e-mail: dferg1000@aol.com

Index

About the Author

Dan Ferguson is the author of two earlier books, *Killer Car Stereo on a Budget* and *Ultimate Car Audio,* and several articles for such periodicals as *Speakerbuilder.* An avid speaker builder all his adult life, he is a mechanical engineer with a Master of Science degree from Clemson University. He practiced engineering for several years before entering management, and has worked at both DuPont and Kimberly-Clark.

CPSIA information can be obtained at www.ICGtesting.com
Printed in the USA
LVOW02s1542311013

359481LV00012B/494/P